The Open University

M33

B3

Classification of surfaces

This publication forms part of an Open University course. Details of this and other Open University courses can be obtained from the Student Registration and Enquiry Service, The Open University, PO Box 197, Milton Keynes, MK7 6BJ, United Kingdom: tel. +44 (0)870 333 4340, e-mail general-enquiries@open.ac.uk

Alternatively, you may visit the Open University website at http://www.open.ac.uk where you can learn more about the wide range of courses and packs offered at all levels by The Open University.

To purchase a selection of Open University course materials, visit the webshop at www.ouw.co.uk, or contact Open University Worldwide, Michael Young Building, Walton Hall, Milton Keynes, MK7 6AA, United Kingdom, for a brochure: tel. +44 (0)1908 858785, fax +44 (0)1908 858787, e-mail ouwenq@open.ac.uk

The Open University, Walton Hall, Milton Keynes, MK7 6AA.

First published 2006.

Edited, designed and typeset by The Open University, using the Open University T X System.

Printed and bound in the United Kingdom by The Charlesworth Group, Wakefield.

ISBN 0 7492 4131 4

1.1

Contents

Introduction

In this unit we complete the journey we started in *Unit B1* — investigating the geometry of surfaces by using algebra.

In *Unit B1* we introduced the three characteristic numbers for a compact surface — the Euler characteristic χ, the boundary number β and the orientability number ω — and we stated the Classification Theorem for compact surfaces:

> *two compact surfaces are homeomorphic if and only if they have the same values for the characteristic numbers.*

We already know from *Unit B1* that the characteristic numbers are topological invariants, and hence that if two compact surfaces are homeomorphic then they have the same values for the characteristic numbers. In this unit we prove the converse: if two compact surfaces have the same values for the characteristic numbers then they are homeomorphic. This will complete the proof of the Classification Theorem.

The argument underlying the proof is as follows. Given any compact surface, we choose any one of the (infinitely many) subdivisions that can be drawn on it. This subdivision gives rise to a number of edge equations, and we choose any one of these. Different choices of subdivision lead to different edge equations, but all edge equations arising in this way give rise to the same characteristic numbers — those of the original surface.

Edge equations were introduced in *Unit B2*, Section 3.

We shall see that each edge equation can be rearranged into a unique standard, or *canonical*, form: this rearrangement changes the edge equation, and therefore the subdivision of the surface, but gives a homeomorphism of the surface. It follows that every edge equation associated with a given surface reduces to the *same* canonical form. Moreover, the characteristic numbers of the surface can easily be read off from the canonical form.

Suppose now that we have two compact surfaces with the same characteristic numbers. The above process associates with each surface an edge equation with the same characteristic numbers as the surface. Because there is a *unique* canonical form corresponding to each admissible set of characteristic numbers, the two surfaces give rise to the same canonical form. But each surface is homeomorphic to the surface that corresponds to the canonical form of the edge equation, and so the two surfaces are homeomorphic, as required. The proof appears in Section 3; some preliminary work on edge equations is undertaken in Section 2.

In Section 4, we meet a new way of constructing new surfaces from old ones — the *connected sum* construction: this is a way of gluing two surfaces together to obtain another surface. We show that the connected sum construction leads to a useful notation for surfaces, and we explain how to calculate the characteristic numbers of a compact surface given in this form. We conclude the unit by classifying all compact surfaces in terms of connected sums of simple compact surfaces.

Study guide

Section 1, *Operations on edge equations*, is a short section. We introduce cut-and-glue operation on surfaces, and show that they are homeomorphisms of the surface: they change the subdivision on the surface, and therefore its edge equation, but do not change the homeomorphism class of the surface. You should not spend too long on this section.

In Section 2, *Rearranging edge equations*, we further shift our emphasis towards working algebraically with the edge equations, rather than geometrically with the more cumbersome polygons with edge identifications. We present six lemmas that allow you to rearrange any edge equation; you should make sure that you understand them before proceeding to Section 3.

Section 3, *The Classification Theorem*, is the most important section of the unit. Using the results of Section 2, we show that any edge equation can be reduced to a unique canonical form, and we describe a systematic approach for doing so. We then use the existence of this canonical form to prove the Classification Theorem. We recommend that you read this section twice — once without the proofs (to familiarize yourself with the main ideas), and then again with the proofs.

In Section 4, *Connected sums of surfaces*, we define the connected sum of surfaces. We introduce an elegant notation for handling the construction, and explain how this notation relates to the canonical form of an edge equation. We also show how to classify compact surfaces in terms of connected sums. This gives us a geometric interpretation of the Classification Theorem.

There is no software associated with this unit.

1 Operations on edge equations

After working through this section, you should be able to:

- ▶ transform a given edge equation of a surface by a sequence of *cut-and-glue operations*;
- ▶ use the *block notation* for edge equations;
- ▶ appreciate that cut-and-glue operations are homeomorphisms of the corresponding surface;
- ▶ understand what are meant by *equivalent* edge equations;
- ▶ appreciate that surfaces corresponding to equivalent edge equations have the same characteristic numbers.

In this section we introduce *cut-and-glue operations* on surfaces. These preserve the characteristic numbers of the surface, are homeomorphisms of the surface, and enable us to rearrange edge equations for the surface. We also introduce the *block notation* for edge equations.

1.1 Cut-and-glue operations

To carry out a **cut-and-glue operation** on a surface given as a polygon with edge identifications

- we draw an edge in the interior of the polygon that joins two vertices of the polygon, and then cut along the edge;
- we use the edge identifications on the original polygon to glue the piece back on.

The two vertices may correspond to the same vertex on the surface.

Figure 1.1 illustrates a cut-and-glue operation on a pentagon with edge identifications. Here

- we draw the internal edge y and then cut along it, to obtain two polygons;
- we glue the polygons along the identified edges a, to obtain a single polygon again.

Note that here we have to turn the triangle over to carry out the gluing.

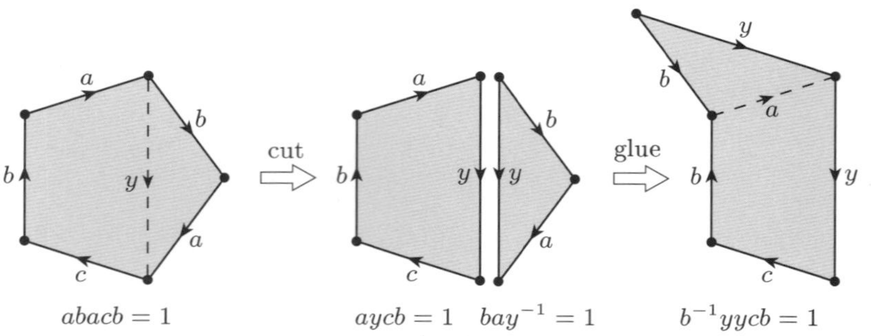

Figure 1.1

Algebraically, in terms of the edge equations,

- we begin with the edge equation $abacb = 1$;
- after cutting along the edge y, we obtain the two edge equations

$$aycb = 1 \quad \text{and} \quad bay^{-1} = 1,$$

  the second of which is equivalent to the edge equation $a = b^{-1}y$;
- after gluing along the edge a (that is, substituting for a in the first equation), we obtain the edge equation $b^{-1}yycb = 1$.

You saw how to manipulate edge equations like this in Subsection 3.3 of *Unit B2*.

In general, we can perform cut-and-glue operations algebraically, using edge equations, without having to draw or cut up the corresponding polygons — this makes things considerably simpler.

Problem 1.1

Consider the polygon with edge equation $ac^{-1}ddf^{-1}f^{-1}ab^{-1} = 1$, and cut it along the edge y, as shown in Figure 1.2. Write down the edge equations of the two polygons obtained in this way, and hence obtain the edge equation of the single polygon obtained by gluing these polygons along the identified edges a.

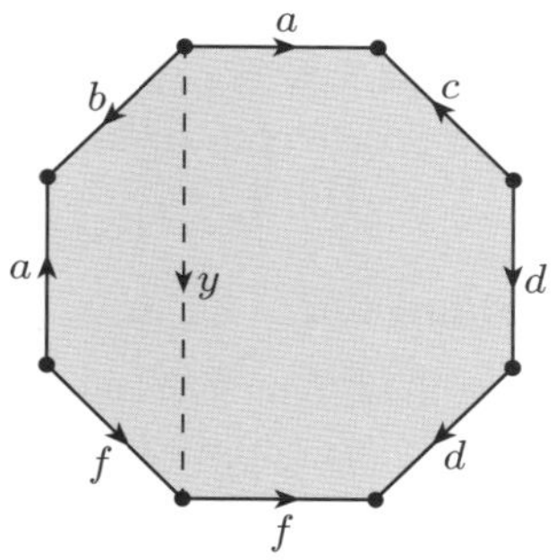

Figure 1.2

1.2 Block notation

One effect of the cut-and-glue operations on edge equations is to introduce new symbols for new edges. The operations also move blocks of symbols around, as we now illustrate.

Consider the edge equation

$$abcb^{-1}xcda^{-1}xdefe^{-1} = 1. \tag{1.1}$$

Suppose that we introduce a cut y from the beginning to the end of the block of symbols $xcda^{-1}$, as shown in Figure 1.3. We obtain two polygons with edge equations

$$abcb^{-1}yxdefe^{-1} = 1 \quad \text{and} \quad xcda^{-1}y^{-1} = 1.$$

From the second equation we deduce the edge equation $x = yad^{-1}c^{-1}$.

Let us now glue along the edge x.

Substituting $x = yad^{-1}c^{-1}$ into the first equation $abcb^{-1}yxdefe^{-1} = 1$, we obtain

$$abcb^{-1}yyad^{-1}c^{-1}defe^{-1} = 1. \tag{1.2}$$

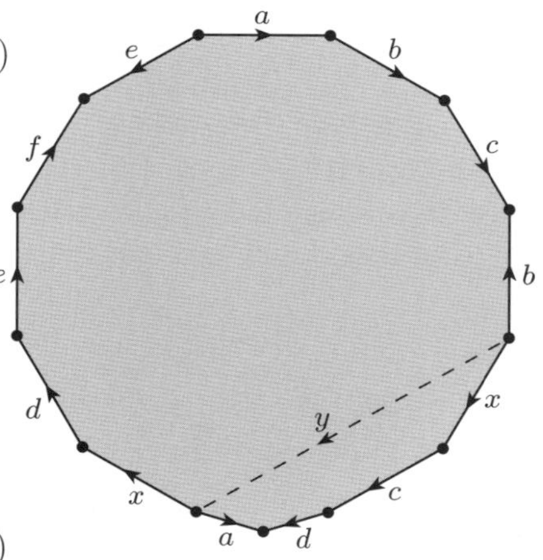

Figure 1.3

Let us compare the original and the final edge equations (1.1) and (1.2).

Each starts with the block $abcb^{-1}$ and ends with the block $defe^{-1}$.

The original edge equation (1.1) has two appearances of the symbol x separated by the block cda^{-1}, while the final edge equation (1.2) has two appearances of the edge y, side by side, followed by the block $ad^{-1}c^{-1}$. Notice what has happened to the block cda^{-1}: in the final edge equation, it is replaced by its inverse $ad^{-1}c^{-1}$.

You will see in Section 3 that replacing separated appearances of the same edge by adjacent appearances of the same edge is the key to the process of simplifying edge equations to canonical form.

In order to study cut-and-glue operations systematically, we introduce some notation for blocks of symbols. Our convention is to use bold-face capital letters ($\boldsymbol{A}$, $\boldsymbol{B}$, $\boldsymbol{C}$, ...) for blocks of symbols. This notation may also be used for blocks consisting of a single letter. We even allow blocks consisting of no edges at all!

Let us try out this notation on our example. We write the block $abcb^{-1}$ as $\boldsymbol{A}$, the block cda^{-1} as $\boldsymbol{B}$, and the block $defe^{-1}$ as $\boldsymbol{C}$. It is also natural to denote $ad^{-1}c^{-1}$ by $\boldsymbol{B}^{-1}$. The original edge equation (1.1) then appears as

$$\boldsymbol{A}x\boldsymbol{B}x\boldsymbol{C} = 1,$$

and the final edge equation (1.2) appears as

$$\boldsymbol{A}yy\boldsymbol{B}^{-1}\boldsymbol{C} = 1.$$

On diagrams, we often draw single edges with a straight line and a block of edges with a wiggly line. In block form, Figure 1.3 now appears as Figure 1.4.

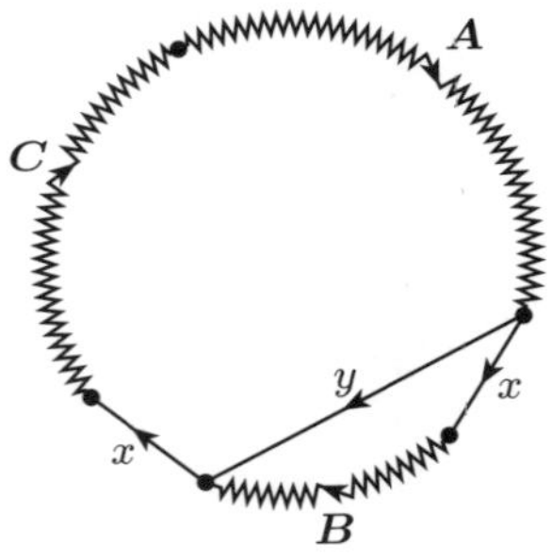

Figure 1.4

Note that any block $\boldsymbol{A}$ has an inverse block $\boldsymbol{A}^{-1}$ in which the symbols appear in reverse order with their orientations reversed. Also, $\boldsymbol{A}\boldsymbol{A}^{-1} = \boldsymbol{A}^{-1}\boldsymbol{A} = 1$.

Problem 1.2

For each of the following blocks $\boldsymbol{A}$, write down $\boldsymbol{A}^{-1}$.

(a) $\boldsymbol{A} = aba^{-1}cdc^{-1}b$

(b) $\boldsymbol{A} = bcab^{-1}c^{-1}a$

1.3 Two particular cut-and-glue operations

In this subsection we look at two cut-and-glue operations that will prove important in the proof of the Classification Theorem.

Suppose we have a polygon with edge identifications in which the edge x appears twice. For the time being we assume that the two appearances of x are in the same sense, so that without loss of generality the corresponding edge equation is of the form $\ldots x \ldots x \ldots = 1$. Let us now draw a new edge y that divides the polygon into two parts, starting at the start of one of the edges x, and each part containing an occurrence of x, as illustrated in Figure 1.5. If we cut along y, we obtain two polygons, each with an edge x and each with an edge y. We now glue the edges x together to obtain a new polygon with edge identifications.

We can change the directions of the arrows on both edges x if necessary.

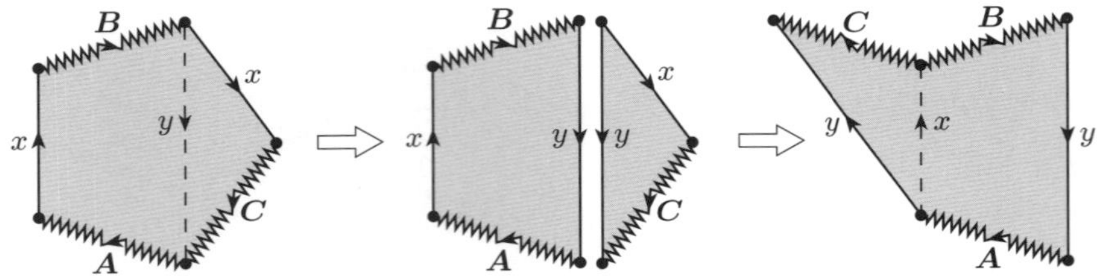

Figure 1.5

Using block notation, we may write the edge equation of the original polygon as $\boldsymbol{A}x\boldsymbol{B}x\boldsymbol{C} = 1$, and of the two polygons as $x\boldsymbol{C}y^{-1} = 1$ and $\boldsymbol{A}x\boldsymbol{B}y = 1$. Each of these equations defines the new edge y, and we may then rewrite them as $y = x\boldsymbol{C}$ and $y = \boldsymbol{B}^{-1}x^{-1}\boldsymbol{A}^{-1}$. Also, since $x = y\boldsymbol{C}^{-1}$ and $x\boldsymbol{C} = y$, we may rewrite the original edge equation in terms of y, as $\boldsymbol{A}y\boldsymbol{C}^{-1}\boldsymbol{B}y = 1$, which is the edge equation of the new polygon.

Since x already appears in $\boldsymbol{A}x\boldsymbol{B}x\boldsymbol{C} = 1$, we know that it cannot also appear in $\boldsymbol{A}$, $\boldsymbol{B}$ or $\boldsymbol{C}$.

The cut-and-glue operation we have performed is specified most easily by the equation $y = x\boldsymbol{C}$, and we call this a *cut-and-glue operation defined by the equation* $y = x\boldsymbol{C}$. It can be performed on any polygon with edge identifications whose corresponding edge equation can be written in the form $\boldsymbol{A}x\boldsymbol{B}x\boldsymbol{C} = 1$.

Here any or all of the blocks $\boldsymbol{A}$, $\boldsymbol{B}$ and $\boldsymbol{C}$ may be empty.

Now suppose that the two appearances of the edge x are in opposite senses. Without loss of generality, the corresponding edge equation can be written in the form $\boldsymbol{A}x^{-1}\boldsymbol{B}x\boldsymbol{C} = 1$. Proceeding as before, we cut to obtain two polygons with edge equations $x\boldsymbol{C}y^{-1} = 1$ and $\boldsymbol{A}x^{-1}\boldsymbol{B}y = 1$, and then glue to obtain a new polygon with edge equation $\boldsymbol{A}\boldsymbol{C}y^{-1}\boldsymbol{B}y = 1$. Thus the cut-and-glue operation defined by $y = x\boldsymbol{C}$ can also be performed on any polygon with edge identifications whose corresponding edge equation can be written in the form $\boldsymbol{A}x^{-1}\boldsymbol{B}x\boldsymbol{C} = 1$, where any or all of the blocks $\boldsymbol{A}$, $\boldsymbol{B}$ and $\boldsymbol{C}$ may be empty.

Again, we can change the directions of the arrows on both edges x if necessary.

Many cut-and-glue operations can be defined by an equation $y = x\boldsymbol{C}$, for suitable choices of x and $\boldsymbol{C}$. For example, the operation illustrated in Figure 1.1 is defined by $y = ba$, where the edge x is b and the block $\boldsymbol{C}$ consists of the single edge a.

Problem 1.3

Use block notation to describe the cut-and-glue operation in Problem 1.1.

There is another simple cut-and-glue operation that we shall need. It is different from the cut-and-glue operations described so far in that it involves only gluing, though for convenience we shall still refer to it as a cut-and-glue operation. It allows us to eliminate the edge x from an edge equation of the form $\boldsymbol{A}xx^{-1}\boldsymbol{B} = 1$ to obtain the edge equation $\boldsymbol{A}\boldsymbol{B} = 1$. The corresponding polygons are shown in Figure 1.6.

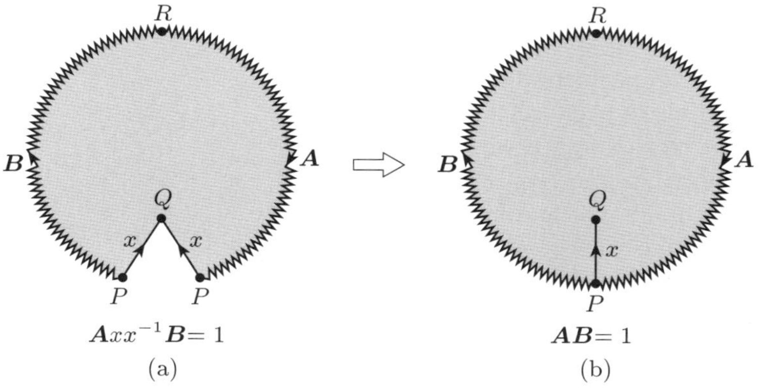

Figure 1.6

In Figure 1.6(a), block $\boldsymbol{A}$ runs from the vertex R to the vertex P, the edge x joins P to the vertex Q, and block $\boldsymbol{B}$ runs from P to R. No vertex in block $\boldsymbol{A}$ or block $\boldsymbol{B}$ can be identified with Q because an edge in one of these blocks would then have to be identified with an edge marked x, which is impossible since the other edge x has already been exhibited and is in neither block. On identifying the two edges marked x, we obtain Figure 1.6(b). The edge x and the vertex Q are now irrelevant and can be dropped, giving the edge equation $\boldsymbol{A}\boldsymbol{B} = 1$.

1.4 Equivalent edge equations

We now explain what we mean by equivalent edge equations and show that various operations on edge equations lead to equivalent edge equations.

> ***Definition***
>
> Two edge equations are **equivalent** if they correspond to homeomorphic surfaces.

We saw in *Unit B2* that edge expressions are equivalent if they correspond to the same polygon with edge identifications. We also saw that, in constructing an edge expression, the starting edge, the choice of edge labels, the directions of edge labels (provided relative senses of identical edges are maintained) and the choice of direction to go round the polygon are all arbitrary. We can easily extend these ideas to edge equations, giving the following equivalent edge equations.

Unit B2, Subsection 3.1.

- The arbitrariness of the starting edge means that all edge equations in which the letters are permuted cyclically are equivalent. Thus an edge equation of the form $\boldsymbol{ABC} = 1$ is equivalent to $\boldsymbol{BCA} = 1$ and $\boldsymbol{CAB} = 1$, though not generally to $\boldsymbol{CBA} = 1$, for example. Similarly $\boldsymbol{A}x\boldsymbol{B}x\boldsymbol{C} = 1$ is equivalent to $x\boldsymbol{B}x\boldsymbol{CA} = 1$.

  This process is often referred to as *cycling the letters*, or simply as *cycling*.

- The arbitrariness of the choice of edge labels means that, for example, $\boldsymbol{A}x\boldsymbol{B}x\boldsymbol{C} = 1$ and $\boldsymbol{A}y\boldsymbol{B}y\boldsymbol{C} = 1$ are equivalent, provided that neither x nor y appears in blocks $\boldsymbol{A}$, $\boldsymbol{B}$ or $\boldsymbol{C}$.

  This is called *relabelling* x as y.

- Provided that the relative senses of identified edges are maintained, the arbitrariness of the directions of the edge labels means that, for example, $\boldsymbol{A}x\boldsymbol{B}x\boldsymbol{C} = 1$ is equivalent to $\boldsymbol{A}x^{-1}\boldsymbol{B}x^{-1}\boldsymbol{C} = 1$, and $\boldsymbol{A}x\boldsymbol{B}x^{-1}\boldsymbol{C} = 1$ is equivalent to $\boldsymbol{A}x^{-1}\boldsymbol{B}x\boldsymbol{C} = 1$.

  We can also replace a string of letters, provided none appears more than once, by a single letter.

- The arbitrariness of the choice of direction to go round the polygon means that, for example, $\boldsymbol{ABC} = 1$ is equivalent to $\boldsymbol{C}^{-1}\boldsymbol{B}^{-1}\boldsymbol{A}^{-1} = 1$ and $\boldsymbol{A}x\boldsymbol{B}x\boldsymbol{C} = 1$ is equivalent to $\boldsymbol{C}^{-1}x^{-1}\boldsymbol{B}^{-1}x^{-1}\boldsymbol{A}^{-1} = 1$.

We also obtain equivalent edge equations if we perform a cut-and-glue operation as described in Subsection 1.1. This is because, when we perform such an operation, all we are effectively doing is drawing one new edge on the surface (the one we cut) and erasing another (the one we glue). We deduce that such a cut-and-glue operation is a homeomorphism, and hence that the edge equations of the surface before and after the operation are equivalent. Thus, for the example in Figure 1.1, the edge equations $abacb = 1$ and $b^{-1}yycb = 1$ are equivalent, and for the example in Subsection 1.2, the edge equations $\boldsymbol{A}x\boldsymbol{B}x\boldsymbol{C} = 1$ and $\boldsymbol{A}yy\boldsymbol{B}^{-1}\boldsymbol{C} = 1$ are equivalent. In particular, a cut-and-glue operation defined by an equation of the form $y = x\boldsymbol{C}$ produces an edge equation equivalent to the original one.

In a similar fashion, we can deduce that the cut-and-glue operation in which $\boldsymbol{A}xx^{-1}\boldsymbol{B} = 1$ is reduced to $\boldsymbol{AB} = 1$ is a homeomorphism, and hence the two edge equations are equivalent.

This is called cancelling the xs.

Problem 1.4

Show that the following pairs of edge equations are equivalent:

(a) $\boldsymbol{A}x\boldsymbol{BC}x\boldsymbol{D} = 1$ and $\boldsymbol{A}y\boldsymbol{C}y\boldsymbol{B}^{-1}\boldsymbol{D} = 1$;

(b) $\boldsymbol{ABC} = 1$ and $\boldsymbol{AB}x^{-1}x\boldsymbol{C} = 1$.

We know from *Unit B1* that homeomorphic surfaces have the same characteristic numbers. We know, by definition, that equivalent edge equations correspond to homeomorphic surfaces. We deduce that surfaces corresponding to equivalent edge equations have the same characteristic numbers. We thus have the following result.

Theorem 1.1

If two compact surfaces have subdivisions with equivalent edge equations, then the surfaces have the same characteristic numbers.

This result will prove useful when we prove the Classification Theorem in Section 3. First, however, we need to prove a number of preliminary results.

2 Rearranging edge equations

After working through this section, you should be able to:

- use the *Moving Lemma*, the *Useful Moving Lemma*, the *Switching Lemma*, the *Jumping Lemma*, the *Assembling Lemma*, and the *Replacement Lemma* to rearrange edge equations to equivalent forms.

We are now ready to treat the analysis of surfaces in a strictly algebraic manner.

As we saw in the previous section, any operation of the form $y = x\boldsymbol{C}$ replaces one subdivision on a surface by another with the same characteristic numbers. Using this operation we can develop a number of standard algebraic moves that combine to give a routine process for simplifying a given edge equation to its equivalent canonical form. In this section we establish these as a list of lemmas, and in Section 3 we use them to obtain the various canonical forms.

Recall that the operation $y = x\boldsymbol{C}$ can be applied to any edge equation that can be written in the form $\boldsymbol{A}x\boldsymbol{B}x\boldsymbol{C} = 1$, and involves replacing $x\boldsymbol{C}$ by y and x by $y\boldsymbol{C}^{-1}$ to obtain $\boldsymbol{A}y\boldsymbol{C}^{-1}\boldsymbol{B}y = 1$. It can also be applied to any edge equation that can be written in the form $\boldsymbol{A}x^{-1}\boldsymbol{B}x\boldsymbol{C} = 1$, in which case it involves replacing $x\boldsymbol{C}$ by y and x^{-1} by $\boldsymbol{C}y^{-1}$ to obtain $\boldsymbol{A}\boldsymbol{C}y^{-1}\boldsymbol{B}y = 1$.

Recall also that we get equivalent edge equations irrespective of which way we go round a polygon with edge identifications. This means that $\boldsymbol{A}x\boldsymbol{B}x\boldsymbol{C} = 1$ is equivalent to $\boldsymbol{C}^{-1}x^{-1}\boldsymbol{B}^{-1}x^{-1}\boldsymbol{A}^{-1} = 1$. Since relabelling produces an equivalent edge equation, we can write $\boldsymbol{D} = \boldsymbol{C}^{-1}$, $\boldsymbol{E} = \boldsymbol{B}^{-1}$, $\boldsymbol{F} = \boldsymbol{A}^{-1}$ and $u = x^{-1}$ to obtain $\boldsymbol{D}u\boldsymbol{E}u\boldsymbol{F} = 1$. Applying the operation $y = x\boldsymbol{C}$ to $\boldsymbol{A}x\boldsymbol{B}x\boldsymbol{C} = 1$ is thus equivalent to applying $v = \boldsymbol{D}u$ to $\boldsymbol{D}u\boldsymbol{E}u\boldsymbol{F} = 1$. It involves replacing $\boldsymbol{D}u$ by v and u by $\boldsymbol{D}^{-1}v$, to obtain $v\boldsymbol{E}\boldsymbol{D}^{-1}v\boldsymbol{F} = 1$. A similar argument applies to edge equations of the form $\boldsymbol{A}x^{-1}\boldsymbol{B}x\boldsymbol{C} = 1$.

We shall often want to apply operations of the form $v = \boldsymbol{D}u$ as well as those of form $y = x\boldsymbol{C}$ in what follows. We shall also need to make frequent use of cycling and relabelling to show that the edge equations to which we apply these operations are of an appropriate form. We shall make considerable use of the fact that cycling, relabelling, changing the directions of edge labels (while respecting relative senses) and changing the direction around a polygon with edge identifications all result in equivalent edge equations. We shall also drop expressions of the form xx^{-1}, or equivalently $x^{-1}x$, from an edge equation. Also since $(x^{-1})^{-1} = x$, any transformation of an expression $\boldsymbol{A}x\boldsymbol{B}x^{-1}\boldsymbol{C}$ applies equally well to an expression $\boldsymbol{A}x^{-1}\boldsymbol{B}x\boldsymbol{C}$.

See Subsection 1.4.

See Subsection 1.3.

Bearing all this in mind, the first lemma we prove is called the *Moving Lemma* (abbreviated to ML), because it enables us to move a block of letters, such as $\boldsymbol{B}$ or $\boldsymbol{C}$, out of a bigger block between two appearances of the letter x.

Lemma 2.1 Moving Lemma (ML)

The following pairs of edge equations are equivalent:

(a) $\boldsymbol{A}x\boldsymbol{BC}x\boldsymbol{D} = 1$ and $\boldsymbol{A}x\boldsymbol{C}x\boldsymbol{B}^{-1}\boldsymbol{D} = 1$;

(b) $\boldsymbol{A}x\boldsymbol{BC}x\boldsymbol{D} = 1$ and $\boldsymbol{AC}^{-1}x\boldsymbol{B}x\boldsymbol{D} = 1$.

In (a), $\boldsymbol{B}$ is moved out of the block $x\boldsymbol{BC}x$.

In (b), $\boldsymbol{C}$ is moved out of the block $x\boldsymbol{BC}x$.

Proof

(a) Put $y = x\boldsymbol{B}$, so $x = y\boldsymbol{B}^{-1}$.

Applying the operation $y = x\boldsymbol{B}$ to the edge equation $\boldsymbol{A}x\boldsymbol{BC}x\boldsymbol{D} = 1$ gives

$$\boldsymbol{A}y\boldsymbol{C}y\boldsymbol{B}^{-1}\boldsymbol{D} = 1.$$

Having eliminated the label x, we can now relabel y as x. Doing so, we obtain the edge equation

$$\boldsymbol{A}x\boldsymbol{C}x\boldsymbol{B}^{-1}\boldsymbol{D} = 1.$$

Note how the symbol x refers to different edges before and after the use of the Moving Lemma.

(b) Put $y = \boldsymbol{C}x$, so $x = \boldsymbol{C}^{-1}y$.

Applying the operation $y = \boldsymbol{C}x$ to the edge equation $\boldsymbol{A}x\boldsymbol{BC}x\boldsymbol{D} = 1$ gives

$$\boldsymbol{AC}^{-1}y\boldsymbol{B}y\boldsymbol{D} = 1.$$

Relabelling y as x, we obtain the edge equation

$$\boldsymbol{AC}^{-1}x\boldsymbol{B}x\boldsymbol{D} = 1. \quad \blacksquare$$

Remarks

(i) In (a), cycling transforms $\boldsymbol{A}x\boldsymbol{BC}x\boldsymbol{D} = 1$ into $\boldsymbol{C}x\boldsymbol{DA}x\boldsymbol{B} = 1$, which is of the form $\boldsymbol{A}x\boldsymbol{B}x\boldsymbol{C} = 1$ with $\boldsymbol{A}$ replaced by $\boldsymbol{C}$, $\boldsymbol{B}$ replaced by $\boldsymbol{DA}$ and $\boldsymbol{C}$ replaced by $\boldsymbol{B}$.

So the operation $y = x\boldsymbol{C}$ applied to $\boldsymbol{A}x\boldsymbol{B}x\boldsymbol{C} = 1$ is equivalent to the operation $y = x\boldsymbol{B}$ applied to $\boldsymbol{C}x\boldsymbol{DA}x\boldsymbol{B} = 1$, which in turn is equivalent to $y = x\boldsymbol{B}$ applied to $\boldsymbol{A}x\boldsymbol{BC}x\boldsymbol{D} = 1$.

(ii) In (b), we apply an operation of the form $v = \boldsymbol{D}u$. Again, cycling and relabelling show that we can legitimately apply this operation.

(iii) The relabelling at the end of each part is legitimate because x cannot appear in any of the blocks $\boldsymbol{A}$, $\boldsymbol{B}$, $\boldsymbol{C}$ or $\boldsymbol{D}$, since otherwise it would have appeared more than twice in the original edge equation.

Problem 2.1

Use the Moving Lemma to show that the following pairs of edge equations are equivalent:

(a) $xyxy = 1$ and $xx = 1$;

(b) $wxyzy^{-1}w = 1$ and $wwyz^{-1}y^{-1}x^{-1} = 1$.

Recall that each equation in part (a) corresponds to the projective plane. In fact, relabelling xy as x in the first equation yields the equation $xx = 1$.

The Moving Lemma has an important consequence, called the *Useful Moving Lemma* (UML); it plays a major role in the proof of the Classification Theorem and in the analysis of many edge equations.

Lemma 2.2 Useful Moving Lemma (UML)

The following edge equations are all equivalent:

$$\boldsymbol{AB}xx\boldsymbol{C} = 1, \quad \boldsymbol{A}x\boldsymbol{B}^{-1}x\boldsymbol{C} = 1 \quad \text{and} \quad \boldsymbol{A}xx\boldsymbol{BC} = 1.$$

Proof We use the Moving Lemma, starting with the middle equation.

We first take $\boldsymbol{B}$ as the empty block, replace $\boldsymbol{C}$ by $\boldsymbol{B}^{-1}$ and $\boldsymbol{D}$ by $\boldsymbol{C}$ (in ML, part (b)), to obtain:

$$\boldsymbol{A}x\boldsymbol{B}^{-1}x\boldsymbol{C} = 1 \Longrightarrow \boldsymbol{AB}xx\boldsymbol{C} = 1.$$

We next replace $\boldsymbol{B}$ by $\boldsymbol{B}^{-1}$, take $\boldsymbol{C}$ as the empty block and replace $\boldsymbol{D}$ by $\boldsymbol{C}$ (in ML, part (a)), to obtain:

In each case, $\boldsymbol{B}^{-1}$ is moved out of the block $x\boldsymbol{B}^{-1}x$.

$$\boldsymbol{A}x\boldsymbol{B}^{-1}x\boldsymbol{C} = 1 \Longrightarrow \boldsymbol{A}xx\boldsymbol{BC} = 1 \quad .$$ ■

Problem 2.2

Show that the following pairs of edge equations are equivalent:

(a) $xx\boldsymbol{DE} = 1$ and $xx\boldsymbol{ED} = 1$;

(b) $auvuxxab = 1$ and $axxuvuab = 1$.

Problem 2.3

The Useful Moving Lemma does not imply that the edge equations $\boldsymbol{AB}x\boldsymbol{C}x\boldsymbol{D} = 1$ and $\boldsymbol{A}x\boldsymbol{C}x\boldsymbol{BD} = 1$ are equivalent. Illustrate this by applying the method of inserting vertices to the edge equations $abxcxb^{-1}d = 1$ and $axcxd = 1$ to determine the corresponding Euler characteristics.

Problem 2.4

Show that the edge equations

$$axxuvuab = 1 \quad \text{and} \quad xxaauuv^{-1}b^{-1} = 1$$

are equivalent.

We will later use the Useful Moving Lemma to bring all pairs of repeated symbols xx to the front of an edge equation, as in Problem 2.4.

The *Switching Lemma* (SL) allows us to switch the order of two blocks $\boldsymbol{B}$ and $\boldsymbol{C}$ between the letters x and x^{-1}.

Lemma 2.3 Switching Lemma (SL)

The following edge equations are equivalent:

$$\boldsymbol{A}x\boldsymbol{BC}x^{-1}\boldsymbol{D} = 1 \quad \text{and} \quad \boldsymbol{A}x\boldsymbol{CB}x^{-1}\boldsymbol{D} = 1.$$

Proof Put $y = x\boldsymbol{B}$, so that $x = y\boldsymbol{B}^{-1}$ and $x^{-1} = \boldsymbol{B}y^{-1}$.

Applying the operation $y = x\boldsymbol{B}$ to the edge equation $\boldsymbol{A}x\boldsymbol{BC}x^{-1}\boldsymbol{D} = 1$ gives

$$\boldsymbol{A}y\boldsymbol{CB}y^{-1}\boldsymbol{D} = 1.$$

Relabelling y as x, we obtain the edge equation

$$\boldsymbol{A}x\boldsymbol{CB}x^{-1}\boldsymbol{D} = 1.$$

■

Remarks

(i) Again, cycling shows that $\boldsymbol{A}x\boldsymbol{BC}x^{-1}\boldsymbol{D} = 1$ is in an appropriate form to apply the operation $y = x\boldsymbol{B}$.

(ii) Changing the directions of the x-labels in the Switching Lemma shows that the following edge equations are also equivalent:

$$\boldsymbol{A}x^{-1}\boldsymbol{BC}x\boldsymbol{D} = 1 \quad \text{and} \quad \boldsymbol{A}x^{-1}\boldsymbol{CB}x\boldsymbol{D} = 1.$$

Our fourth lemma is called the *Jumping Lemma* (JL), because the block $\boldsymbol{B}$ 'jumps over' the block $x\boldsymbol{C}x^{-1}$.

Lemma 2.4 Jumping Lemma (JL)

The following edge equations are equivalent:

$$\boldsymbol{AB}x\boldsymbol{C}x^{-1}\boldsymbol{D} = 1 \quad \text{and} \quad \boldsymbol{A}x\boldsymbol{C}x^{-1}\boldsymbol{BD} = 1.$$

Proof Put $y = \boldsymbol{B}x$, so that $x = \boldsymbol{B}^{-1}y$ and $x^{-1} = y^{-1}\boldsymbol{B}$.

Applying the operation $y = \boldsymbol{B}x$ to the edge equation $\boldsymbol{AB}x\boldsymbol{C}x^{-1}\boldsymbol{D} = 1$ gives

$$\boldsymbol{A}y\boldsymbol{C}y^{-1}\boldsymbol{BD} = 1.$$

Relabelling y as x, we obtain the edge equation

$$\boldsymbol{A}x\boldsymbol{C}x^{-1}\boldsymbol{BD} = 1.$$

■

Remarks

(i) Cycling and changing the directions of the x-labels shows that $\boldsymbol{AB}x\boldsymbol{C}x^{-1}\boldsymbol{D} = 1$ is in an appropriate form to apply the operation $y = \boldsymbol{B}x$.

(ii) Changing the directions of x-labels in the Jumping Lemma shows that the following edge equations are also equivalent:

$$\boldsymbol{AB}x^{-1}\boldsymbol{C}x\boldsymbol{D} = 1 \quad \text{and} \quad \boldsymbol{A}x^{-1}\boldsymbol{C}x\boldsymbol{BD} = 1.$$

Problem 2.5

Use the Jumping Lemma to show that the edge equations

$$\boldsymbol{D}u\boldsymbol{B}u^{-1}x\boldsymbol{C}x^{-1} = 1 \quad \text{and} \quad \boldsymbol{D}x\boldsymbol{C}x^{-1}u\boldsymbol{B}u^{-1} = 1$$

are equivalent.

Just as the Useful Moving Lemma collects pairs xx, so the Assembling Lemma (AL) gathers blocks of the form $xyx^{-1}y^{-1}$. In particular, the Assembling Lemma enables us to rewrite edge equations of the form $\boldsymbol{AB}x\boldsymbol{C}y\boldsymbol{D}x^{-1}\boldsymbol{E}y^{-1}\boldsymbol{F}=1$ as $\boldsymbol{A}xyx^{-1}y^{-1}\boldsymbol{BEDCF}=1$, in which the symbols x and y are 'assembled' into a block of the form $xyx^{-1}y^{-1}$.

This will be useful in the next section.

Lemma 2.5 Assembling Lemma (AL)

The following edge equations are equivalent:

$$\boldsymbol{AB}x\boldsymbol{C}y\boldsymbol{D}x^{-1}\boldsymbol{E}y^{-1}\boldsymbol{F}=1 \quad \text{and} \quad \boldsymbol{A}xyx^{-1}y^{-1}\boldsymbol{BEDCF}=1.$$

Proof The proof uses the Jumping Lemma and the Switching Lemma.

$$\begin{aligned}
&\boldsymbol{AB}x\boldsymbol{C}y\boldsymbol{D}x^{-1}\boldsymbol{E}y^{-1}\boldsymbol{F}=1 \\
&\quad\to \boldsymbol{A}x\boldsymbol{C}y\boldsymbol{D}x^{-1}\boldsymbol{BE}y^{-1}\boldsymbol{F}=1 \quad (\text{JL}: \boldsymbol{B} \text{ jumps over } x \text{ and } x^{-1}) \\
&\quad\to \boldsymbol{A}xy\boldsymbol{D}x^{-1}\boldsymbol{BE}y^{-1}\boldsymbol{CF}=1 \quad (\text{JL}: \boldsymbol{C} \text{ jumps over } y \text{ and } y^{-1}) \\
&\quad\to \boldsymbol{A}xy\boldsymbol{BED}x^{-1}y^{-1}\boldsymbol{CF}=1 \quad (\text{SL}: \boldsymbol{D}x^{-1} \text{ and } \boldsymbol{BE} \text{ switch between } y \text{ and } y^{-1}) \\
&\quad\to \boldsymbol{A}x\boldsymbol{BED}yx^{-1}y^{-1}\boldsymbol{CF}=1 \quad (\text{SL}: y \text{ and } \boldsymbol{BED} \text{ switch between } x \text{ and } x^{-1}) \\
&\quad\to \boldsymbol{A}xyx^{-1}y^{-1}\boldsymbol{BEDCF}=1 \quad (\text{JL}: \boldsymbol{BED} \text{ jumps over } y \text{ and } y^{-1}). \blacksquare
\end{aligned}$$

Remarks

(i) Changing the directions of the x and/or y edge labels in the Assembling Lemma shows that the following pairs of edge equations are also equivalent:

$$\boldsymbol{AB}x^{-1}\boldsymbol{C}y\boldsymbol{D}x\boldsymbol{E}y^{-1}\boldsymbol{F}=1 \quad \text{and} \quad \boldsymbol{A}x^{-1}yxy^{-1}\boldsymbol{BEDCF}=1;$$

$$\boldsymbol{AB}x\boldsymbol{C}y^{-1}\boldsymbol{D}x^{-1}\boldsymbol{E}y\boldsymbol{F}=1 \quad \text{and} \quad \boldsymbol{A}xy^{-1}x^{-1}y\boldsymbol{BEDCF}=1;$$

$$\boldsymbol{AB}x^{-1}\boldsymbol{C}y^{-1}\boldsymbol{D}x\boldsymbol{E}y\boldsymbol{F}=1 \quad \text{and} \quad \boldsymbol{A}x^{-1}y^{-1}xy\boldsymbol{BEDCF}=1.$$

(ii) Notice that the blocks outside the x, y, x^{-1} and y^{-1} terms ($\boldsymbol{A}$, $\boldsymbol{B}$ and $\boldsymbol{F}$) remain in that order, whereas the ones between these terms ($\boldsymbol{C}$, $\boldsymbol{D}$ and $\boldsymbol{E}$) end up in reverse order.

Problem 2.6

Use the Assembling Lemma to show that the edge equations $\boldsymbol{A}xyx^{-1}y^{-1}\boldsymbol{B} = 1$ and $xyx^{-1}y^{-1}\boldsymbol{AB} = 1$ are equivalent.

We will later use the Assembling Lemma to bring all blocks of the form $xyx^{-1}y^{-1}$ to the front of an edge equation, as in Problem 2.6.

Our final lemma is the *Replacement Lemma* (RL), which allows us, in the presence of zz, to replace $xyx^{-1}y^{-1}$ by $xxyy$.

Lemma 2.6 Replacement Lemma (RL)

The following edge equations are equivalent:

$$\boldsymbol{A}zzxyx^{-1}y^{-1}\boldsymbol{B} = 1 \quad \text{and} \quad \boldsymbol{A}zzxxyy\boldsymbol{B} = 1.$$

Proof The proof uses the Useful Moving Lemma.

$\boldsymbol{A}zzxyx^{-1}y^{-1}\boldsymbol{B} = 1$

$\to \boldsymbol{A}zx^{-1}zyx^{-1}y^{-1}\boldsymbol{B} = 1$ (moving x inside zz)

$\to \boldsymbol{A}xzzyx^{-1}y^{-1}\boldsymbol{B} = 1$ (moving x^{-1} outside zz)

$\to \boldsymbol{A}xzxy^{-1}zy^{-1}\boldsymbol{B} = 1$ (moving yx^{-1} inside zz)

$\to \boldsymbol{A}xxz^{-1}y^{-1}zy^{-1}\boldsymbol{B} = 1$ (moving the first z outside xx)

$\to \boldsymbol{A}xxz^{-1}z^{-1}y^{-1}y^{-1}\boldsymbol{B} = 1$ (moving the second z outside $y^{-1}y^{-1}$)

$\to \boldsymbol{A}zzxxyy\boldsymbol{B} = 1$ (relabelling x as z, z^{-1} as x, and y^{-1} as y).

■

Remark

Changing the directions of the x and/or y edge labels in the Replacement Lemma shows that the following edge equations are also equivalent to $\boldsymbol{A}zzxxyy\boldsymbol{B} = 1$:

$$\boldsymbol{A}zzx^{-1}yxy^{-1}\boldsymbol{B} = 1, \quad \boldsymbol{A}zzxy^{-1}x^{-1}y\boldsymbol{B} = 1$$

$$\text{and} \quad \boldsymbol{A}zzx^{-1}y^{-1}xy\boldsymbol{B} = 1.$$

These six lemmas are the basic tricks of the trade. In the next section we use them to put any edge equation into a canonical form.

For convenience, we restate the six lemmas below, with a summary of what each of them enables us to do.

When using the lemmas in assignments, it is enough to quote each lemma by its name or initials.

The following lemmas assert that edge equations of the stated forms are equivalent:

Moving Lemma (ML)

(a) $\boldsymbol{A}x\boldsymbol{BC}x\boldsymbol{D} = 1$ and $\boldsymbol{A}x\boldsymbol{C}x\boldsymbol{B}^{-1}\boldsymbol{D} = 1$;

(b) $\boldsymbol{A}x\boldsymbol{BC}x\boldsymbol{D} = 1$ and $\boldsymbol{AC}^{-1}x\boldsymbol{B}x\boldsymbol{D} = 1$.

Useful Moving Lemma (UML)

$\boldsymbol{AB}xx\boldsymbol{C} = 1$, $\boldsymbol{A}x\boldsymbol{B}^{-1}x\boldsymbol{C} = 1$ and $\boldsymbol{A}xx\boldsymbol{BC} = 1$.

Switching Lemma (SL)

$\boldsymbol{A}x\boldsymbol{BC}x^{-1}\boldsymbol{D} = 1$ and $\boldsymbol{A}x\boldsymbol{CB}x^{-1}\boldsymbol{D} = 1$.

Jumping Lemma (JL)

$\boldsymbol{AB}x\boldsymbol{C}x^{-1}\boldsymbol{D} = 1$ and $\boldsymbol{A}x\boldsymbol{C}x^{-1}\boldsymbol{BD} = 1$.

Assembling Lemma (AL)

$\boldsymbol{AB}x\boldsymbol{C}y\boldsymbol{D}x^{-1}\boldsymbol{E}y^{-1}\boldsymbol{F} = 1$ and $\boldsymbol{A}xyx^{-1}y^{-1}\boldsymbol{BEDCF} = 1$.

Replacement Lemma (RL)

$\boldsymbol{A}zzxyx^{-1}y^{-1}\boldsymbol{B} = 1$ and $\boldsymbol{A}zzxxyy\boldsymbol{B} = 1$.

Note that only the Moving Lemma, the Useful Moving Lemma and the Replacement Lemma change the signs of the edges in the equation.

The Moving Lemma and the Useful Moving Lemma allow us to bring closer together repeated occurrences of the same symbol appearing in the same sense $(\ldots x \ldots x \ldots)$.

The Switching Lemma and the Jumping Lemma allow us a limited ability to switch the order of blocks that appear between two occurrences of the same symbol appearing in opposite senses $(\ldots x \ldots x^{-1} \ldots)$.

We conclude this section with two problems that give you the chance to decide which of the six lemmas to use in different circumstances.

Problem 2.7

Show that the following pairs of edge equations are equivalent, stating which lemma(s) you use in each case:

(a) $abcacb^{-1} = 1$ and $xxyyzz = 1$;

(b) $aba^{-1}b^{-1}e = 1$ and $aba^{-1}b^{-1}ded^{-1} = 1$.

Problem 2.8

(a) Show that the edge equations $xxaba^{-1}b^{-1} = 1$ and $xx = 1$ are not equivalent.

(b) Use the result of part (a) to show that an edge equation of the form $\boldsymbol{CAB} = 1$ is not necessarily equivalent to one of the form $\boldsymbol{CBA} = 1$.

3 The Classification Theorem

After working through this section, you should be able to:

- explain what is meant by the *canonical form* of an edge equation;
- reduce any given edge equation to canonical form;
- write down the characteristic numbers of any edge equation in canonical form;
- state and use the Classification Theorem for compact surfaces.

In this section we show that, by judicious use of the six lemmas of Section 2, we can write the edge equation of any compact surface in one of three standard forms: these are the *canonical forms*. Using these canonical forms, we prove the Classification Theorem for compact surfaces, which is the main result of Block B.

You are encouraged to skim through this section first, noting the main results but omitting the proofs. Later in the section we present some Worked problems in which we follow the method of the proof.

3.1 Canonical form

There are three canonical forms for edge equations, as listed in the following theorem.

Theorem 3.1 Canonical Form Theorem

Every edge equation can be reduced to exactly one of the following forms:

(a) $aa^{-1} = 1$;

(b) $a_1 b_1 a_1^{-1} b_1^{-1} \cdots a_m b_m a_m^{-1} b_m^{-1} \, x_1 c_1 x_1^{-1} \cdots x_n c_n x_n^{-1} = 1 \ (m \geq 0, n \geq 0)$;

(c) $a_1 a_1 \cdots a_k a_k \, x_1 c_1 x_1^{-1} \cdots x_n c_n x_n^{-1} = 1 \ (k > 0, n \geq 0)$.

The numbers m, n and k are determined by the original edge equation.

Remarks

(i) Each of the above forms is called the **canonical form** of the original edge equation (or, briefly, a **canonical edge equation**).

(ii) Canonical forms (a) and (b) correspond to *orientable* surfaces, since they contain no repeated symbols $\cdots x \cdots x \cdots$: canonical form (a) corresponds to the sphere, while canonical form (b) corresponds to all other orientable surfaces.

(iii) Canonical form (c) corresponds to *non-orientable* surfaces, since it contains at least one pair of repeated symbols in the same sense.

The relationship between repeated edge labels and orientability was discussed in *Unit B2*, Subsection 3.2.

(iv) The canonical form of an edge equation either corresponds to a sphere ($aa^{-1} = 1$) or is made up from:

- blocks of the form xx, which indicate *cross-caps*;
- blocks of the form $aba^{-1}b^{-1}$, which indicate *handles*;
- blocks of the form xcx^{-1}, which indicate *holes*.

We saw, at the end of *Unit B2* Section 3, that an unrepeated letter between two repeated letters corresponds to a hole in a surface.

Remark (iv) suggests that the corresponding surface is made up of cross-caps or handles, with perhaps some open discs removed (to create the boundary components). This is indeed the case, and in Section 4 we shall see how any surface can be built from a sphere by attaching cross-caps or handles and then removing open discs. We deduce that any surface can be thought of as either the sphere, or a certain number of toruses, or a certain number of projective planes, in each case with perhaps some open discs removed.

Problem 3.1

(a) Classify each of the following edge equations as having canonical form (a), (b) or (c), and write down the values of m, n and k, as appropriate:

torus: $aba^{-1}b^{-1} = 1$

2-fold torus: $aba^{-1}b^{-1}cdc^{-1}d^{-1} = 1$

projective plane: $abab = 1$

(b) By using the Useful Moving Lemma, explain how the Klein bottle, with edge equation $abab^{-1} = 1$, fits into this classification.

These edge equations are given in *Unit B2*, at the end of Section 3.

We now prove that every edge equation can be reduced to canonical form: that is, we prove the *existence* of the canonical forms. Later in this subsection we complete the proof of Theorem 3.1 by showing that the numbers m, n and k are determined uniquely by the original edge equation: that is, we prove the *uniqueness* of the canonical forms.

The techniques we introduce in this proof are those you can use to reduce *any* edge equation to canonical form. They can be refined to a computer algorithm.

Proof of existence

Our method proceeds systematically.

If the edge equation is of the form $a_1a_1^{-1}a_2a_2^{-1}\cdots a_qa_q^{-1}$, we cancel all but one pair and after relabelling obtain

$$aa^{-1} = 1,$$

the first of the canonical forms.

We know from Sections 1 and 2 that the edge equations $\boldsymbol{A}x x^{-1}\boldsymbol{B} = 1$ and $\boldsymbol{AB} = 1$ are equivalent.

Otherwise, we reduce the equation in three steps, in which we deal successively with:

- cross-caps: pairs of the form xx;
- handles: expressions of the form $aba^{-1}b^{-1}$;
- holes: expressions of the form xcx^{-1}.

Step 1 Assemble cross-caps

We apply the Useful Moving Lemma, followed by cycling if necessary, to pull all pairs of symbols appearing in the same sense to the front of the edge equation.

See Problem 2.4, where this is done.

The edge equation now has the form

$$a_1a_1\cdots a_ra_r\boldsymbol{W} = 1,$$

Note that none of the symbols $a_1, \ldots, a_r$ can appear in $\boldsymbol{W}$.

where $r = 0$ if no symbols were repeated in the same sense, and where the block $\boldsymbol{W}$ contains no pairs of symbols repeated in the same sense.

We rewrite this equation as

$$\boldsymbol{AW} = 1,$$

where $\boldsymbol{A}$ is the empty block or is of the form $a_1a_1\cdots a_ra_r$ $(r>0)$ and $\boldsymbol{W}$ contains no pairs of symbols repeated in the same sense. If $\boldsymbol{W}$ is empty, we have finished.

Step 2 Assemble handles

We now look to see if $\boldsymbol{W}$ is of the form $\boldsymbol{B}y_1\boldsymbol{C}z_1\boldsymbol{D}y_1^{-1}\boldsymbol{E}z_1^{-1}\boldsymbol{F}$.

If it is, we use the Assembling Lemma to rewrite $\boldsymbol{AW}=1$ in the form

$$\boldsymbol{A}y_1z_1y_1^{-1}z_1^{-1}\boldsymbol{BEDCF}=1.$$

We rewrite this equation as

$$\boldsymbol{A}_1\boldsymbol{W}_1=1,$$

where $\boldsymbol{A}_1=\boldsymbol{A}y_1z_1y_1^{-1}z_1^{-1}$. We now look to see if $\boldsymbol{W}_1$ is in an appropriate form to allow us to apply the Assembling Lemma to $\boldsymbol{A}_1\boldsymbol{W}_1=1$, and if so we apply it to obtain

$$\boldsymbol{A}_2\boldsymbol{W}_2=\boldsymbol{A}_1y_2z_2y_2^{-1}z_2^{-1}\boldsymbol{W}_2=\boldsymbol{A}y_1z_1y_1^{-1}z_1^{-1}y_2z_2y_2^{-1}z_2^{-1}\boldsymbol{W}_2=1.$$

We continue in this way until we can apply the Assembling Lemma no more. The edge equation now has the form

$$\boldsymbol{A}\,y_1z_1y_1^{-1}z_1^{-1}\cdots y_mz_my_m^{-1}z_m^{-1}\,\boldsymbol{X}=1,$$

for some block $\boldsymbol{X}$, where $\boldsymbol{A}$ is the empty block or is of the form $a_1a_1\cdots a_ra_r$, and where $m\geq 0$.

All the repeated symbols (if any) have now been pulled to the front of the edge equation, followed by all the handles (if any), followed by a block $\boldsymbol{X}$ of symbols that is so far resistant to our efforts. If $\boldsymbol{X}$ is empty, we have finished.

We note that use of the Assembling Lemma has not changed the sense of any edge label, so $\boldsymbol{X}$ cannot contain any new pair of symbols in the same sense which we might have to deal with as in Step 1.

Step 3 Assemble holes

In Step 1 we pulled all pairs of repeated symbols occurring in the same sense to the front of the edge equation, and in Step 2 we created no new pairs of repeated symbols in the same sense. So the block $\boldsymbol{X}$ contains no repeated symbols in the same sense. It may, however, contain repeated symbols in opposite senses. We also know, from Step 2, that it contains no handles — that is, nothing of the form $\cdots y\cdots z\cdots y^{-1}\cdots z^{-1}$.

If the symbol x appears twice in the form $\cdots x^{-1}\cdots x\cdots$, we relabel x as x^{-1}, giving $\cdots x\cdots x^{-1}\cdots$. If the block separating x and x^{-1} is empty, we have $\cdots xx^{-1}\cdots$, and we delete both x and x^{-1}. So we can assume that every repeated symbol in $\boldsymbol{X}$ occurs in the form $\cdots x\boldsymbol{D}x^{-1}\cdots$, where the block $\boldsymbol{D}$ is non-empty.

We are therefore interested in edge equations of the form $\boldsymbol{ABX}=1$, where all occurrences of xx are in $\boldsymbol{A}$ and all occurrences of $xyx^{-1}y^{-1}$ are in $\boldsymbol{B}$. So $\boldsymbol{A}$ consists of blocks of symbols of the form (xx followed by a block of cross-caps), and $\boldsymbol{B}$ consists of symbols of the form ($xyx^{-1}y^{-1}$ followed by a block of handles).

Suppose that there is a repeated symbol in $\boldsymbol{X}$, so $\boldsymbol{X}$ has the form

$$\ldots x\ldots x^{-1}\ldots\,.$$

We write this as $\boldsymbol{X}=\boldsymbol{C}(x\boldsymbol{D}x^{-1})\boldsymbol{E}$. We can apply the Jumping Lemma and rewrite the edge equation $\boldsymbol{ABC}(x\boldsymbol{D}x^{-1})\boldsymbol{E}=1$ as $\boldsymbol{AB}(x\boldsymbol{D}x^{-1})\boldsymbol{CE}=1$. When we do this we must be sure that $(x\boldsymbol{D}x^{-1})\boldsymbol{CE}$

contains no cross-caps or handles. It clearly contains no cross-caps, because the Jumping Lemma does not change the sign of any symbol. To see that this use of the Jumping Lemma introduces no handles, we argue as follows.

We must show that if an edge equation has the form $\boldsymbol{ABC}(x\boldsymbol{D}x^{-1})\boldsymbol{E} = 1$, and if $\boldsymbol{C}(x\boldsymbol{D}x^{-1})\boldsymbol{E}$ is not of the form

$$\ldots p \ldots q \ldots p^{-1} \ldots q^{-1} \ldots,$$

then the block $(x\boldsymbol{D}x^{-1})\boldsymbol{CE} = 1$ obtained by using the Jumping Lemma is also not of this form.

Consider the edge equation $\boldsymbol{ABC}(x\boldsymbol{D}x^{-1})\boldsymbol{E} = 1$, and let p be a symbol in $\boldsymbol{D}$. The symbol p^{-1} cannot occur in $\boldsymbol{C}$, since otherwise $\boldsymbol{ABC}(x\boldsymbol{D}x^{-1})\boldsymbol{E}$ would be of the form

$$\ldots p^{-1} \ldots x \ldots p \ldots x^{-1} \ldots,$$

which by our assumption it is not. Similarly, the symbol p^{-1} cannot occur in $\boldsymbol{E}$, since otherwise $\boldsymbol{ABC}(x\boldsymbol{D}x^{-1})\boldsymbol{E}$ would be of the form

$$\ldots x \ldots p \ldots x^{-1} \ldots p^{-1} \ldots,$$

which by our assumption it is not.

So no edge appearing in $\boldsymbol{D}$ is repeated, and we may replace $\boldsymbol{D}$ by a single edge c. Our edge equation now has the form $\boldsymbol{ABC}(xcx^{-1})\boldsymbol{E} = 1$.

Now suppose for a contradiction that $\boldsymbol{C}(xcx^{-1})\boldsymbol{E}$ is not of the form

$$\ldots p \ldots q \ldots p^{-1} \ldots q^{-1} \ldots,$$

but $(xcx^{-1})\boldsymbol{CE}$ has the form

$$\ldots p \ldots q \ldots p^{-1} \ldots q^{-1} \ldots.$$

The symbol x cannot be either p or q, because nothing between p and q is repeated. The symbol c cannot be p or q, because it is not repeated. So the symbols p, q, p^{-1} and q^{-1} in $(xdx^{-1})\boldsymbol{CE}$ do not occur in the block xcx^{-1}. But this means that they occur in the same order in $(xcx^{-1})\boldsymbol{CE}$ as they do in $\boldsymbol{C}(xcx^{-1})\boldsymbol{E}$. We deduce that if $(xcx^{-1})\boldsymbol{CE}$ is of the form

$$\ldots p \ldots q \ldots p^{-1} \ldots q^{-1} \ldots,$$

then so is $\boldsymbol{C}(xcx^{-1})\boldsymbol{E}$, which by our assumption it is not. This is the contradiction we seek: it follows that using the Jumping Lemma in the present context does not introduce a handle.

We now repeat the process until the block $\boldsymbol{X}$ has the form

$$x_1 c_1 x_1^{-1} \ldots x_s c_s x_s^{-1} \boldsymbol{Y},$$

where $\boldsymbol{Y}$ is a block of symbols that occur nowhere else in the edge equation. Note that if $\boldsymbol{X}$ had initially contained no repeated symbols then we would already be at this stage in the analysis. If $\boldsymbol{Y}$ is empty, there is nothing to do. If $\boldsymbol{Y}$ is not empty, then $\boldsymbol{Y}$ can be replaced by a single symbol d.

To complete the process of obtaining a canonical form, we now add the pair of new symbols $y^{-1}y$ to the end of the equation to produce the equivalent equation

$$\boldsymbol{AB} x_1 c_1 x_1^{-1} \cdots x_s c_s x_s^{-1} d y^{-1} y = 1.$$

This process of replacing a symbol d in an edge equation of this form is called *creating a hole* or *making a cuff out of a hole*.

Cycling gives

$$y \boldsymbol{AB} x_1 c_1 x_1^{-1} \cdots x_s c_s x_s^{-1} d y^{-1} = 1.$$

If $\boldsymbol{A}$ is non-empty, several applications of the Useful Moving Lemma give

$$\boldsymbol{A} y \boldsymbol{B} x_1 c_1 x_1^{-1} \cdots x_s c_s x_s^{-1} d y^{-1} = 1.$$

If $\boldsymbol{A}$ is empty, we have this form of the equation without applying the Useful Moving Lemma.

If $\boldsymbol{B}$ is non-empty, several applications of the Assembling Lemma (with $\boldsymbol{C}$, $\boldsymbol{D}$ and $\boldsymbol{E}$ empty) give

$$\boldsymbol{AB}yx_1c_1x_1^{-1}\cdots x_sc_sx_s^{-1}dy^{-1}=1.$$

If $\boldsymbol{B}$ is empty, we have this form of the equation without applying the Assembling Lemma.

If $s\neq 0$, several applications of the Jumping Lemma give

$$\boldsymbol{AB}x_1c_1x_1^{-1}\cdots x_sc_sx_s^{-1}ydy^{-1}=1.$$

If $s=0$, we have this form of the equation without applying the Jumping Lemma.

We have thus reduced the original equation to one which, after some relabelling, has the form

$$a_1a_1\cdots a_ra_ry_1z_1y_1^{-1}z_1^{-1}\cdots y_mz_my_m^{-1}z_m^{-1}x_1c_1x_1\cdots x_nc_nx_n^{-1}=1. \qquad (3.1)$$

We now need a few small finishing touches to complete the proof that every edge equation can be reduced to one of three canonical forms. We have shown that every edge equation can be reduced either to $aa^{-1}=1$ (the first of the canonical forms) or, by using the three steps above, to the form (3.1).

If $r=0$ in (3.1), after some relabelling we obtain

$$a_1b_1a_1^{-1}b_1^{-1}\cdots a_mb_ma_m^{-1}b_m^{-1}x_1c_1x_1^{-1}\cdots x_nc_1x_n^{-1}=1 \quad (m\geq 0, n\geq 0),$$

the third of the canonical forms.

If $r>0$, we apply the Replacement Lemma to (3.1) m times to obtain

$$a_1a_1\cdots a_ra_ry_1y_1z_1z_1\cdots y_my_mz_mz_mx_1c_1x_1^{-1}\cdots x_nc_nx_n^{-1}=1,$$

which on relabelling and putting $k=2m+r$ can be written as

$$a_1a_1\cdots a_ka_kx_1c_1x_1^{-1}\cdots x_nc_nx_n^{-1}=1 \quad (k>0, n\geq 0),$$

the second of the canonical forms.

Thus, every edge equation can be written in one of the three canonical forms. ■

We have thus proved the existence of the canonical forms. It remains to prove their uniqueness, which takes up the remainder of this subsection. Our first step is to calculate the characteristic numbers of the surfaces described by the canonical edge equations.

Characteristic numbers

Orientable surfaces

We saw at the start of this subsection that canonical forms (a) and (b) correspond to orientable surfaces. Form (a) corresponds to a sphere, which has characteristic numbers $\chi=2$, $\beta=0$, $\omega=0$. To find the characteristic numbers associated with surfaces with a canonical edge equation of form (b), we first need a preliminary result.

Lemma 3.2

The method of inserting vertices applied to an edge equation of the form

$$a_1b_1a_1^{-1}b_1^{-1}\cdots a_mb_ma_m^{-1}b_m^{-1}\,x_1c_1x_1^{-1}\cdots x_nc_nx_n^{-1}=1,$$

yields

$$Pa_1Pb_1Pa_1^{-1}Pb_1^{-1}P\cdots Pa_mPb_mPa_m^{-1}Pb_m^{-1}\,Px_1Q_1c_1Q_1x_1^{-1}P \cdots Px_nQ_nc_nQ_nx_n^{-1}P=1.$$

Proof Suppose that the edge a_1 starts at P. Then the process of inserting vertices into the edge equation begins

$$\underline{Pa_1}\, b_1\, \underline{a_1^{-1} P}\, b_1^{-1} \cdots a_m b_m a_m^{-1} b_m^{-1} x_1 c_1 x_1^{-1} \cdots x_n c_n\, \underline{x_n^{-1} P} = 1,$$

showing that the edge b_1 ends at P.

This gives

$$Pa_1\, \underline{b_1 P}\, a_1^{-1} P b_1^{-1} \cdots a_m b_m a_m^{-1} b_m^{-1}\, x_1 c_1 x_1^{-1} \cdots x_n c_n x_n^{-1} P = 1,$$

showing that the edge a_1 ends at P. This gives

$$P\, \underline{a_1 P}\, b_1 P a_1^{-1} P b_1^{-1} \cdots a_m b_m a_m^{-1} b_m^{-1}\, x_1 c_1 x_1^{-1} \cdots x_n c_n x_n^{-1} P = 1,$$

showing that the edge b_1 starts at P. This gives

$$Pa_1 P b_1 P a_1^{-1} P b_1^{-1} P \cdots a_m b_m a_m^{-1} b_m^{-1} x_1 c_1 x_1^{-1} \cdots x_n c_n x_n^{-1} P = 1.$$

It follows that the block $a_2 b_2 a_2^{-1} b_2^{-1}$ starts at P, allowing us to initiate the same process of inserting vertices for this block, yielding

$$\cdots Pa_2 P b_2 P a_2^{-1} P b_2^{-1} P \cdots.$$

Continuing in this way for each of the handles, we obtain

$$Pa_1 P b_1 P a_1^{-1} P b_1^{-1} P \cdots P a_m P b_m P a_m^{-1} P b_m^{-1} P x_1 c_1 x_1^{-1} \cdots x_n c_n x_n^{-1} P = 1.$$

From this we see that x_1 starts at P, so the block $x_1 c_1 x_1^{-1}$ starts and ends at P. This means that x_2 starts at P and hence the block $x_2 c_2 x_2^{-1}$ ends at P. Proceeding in this way, we see that each hole starts and ends at P, giving

$$Pa_1 P b_1 P a_1^{-1} P b_1^{-1} \cdots P a_m P b_m P a_m^{-1} P b_m^{-1} P x_1 c_1 x_1^{-1} P \cdots P x_n c_n x_n^{-1} P = 1.$$

We cannot now insert any more Ps, so we move on to a new vertex Q_1 and insert this at the end of x_1 to give

$$\cdots P x_1 Q_1 c_1 Q_1 x_1^{-1} P \cdots$$

for the first hole. Proceeding in this way for each hole, we obtain

$$Pa_1 P b_1 P a_1^{-1} P b_1^{-1} P \cdots P a_m P b_m P a_m^{-1} P b_m^{-1} P x_1 Q_1 c_1 Q_1 x_1^{-1} P$$
$$\cdots P x_n Q_n c_n Q_n x_n^{-1} P = 1.$$

Thus, there are $n+1$ vertices: $P, Q_1, \ldots, Q_n$. ■

We can now deduce the following theorem.

Theorem 3.3

The characteristic numbers of any surface whose edge equation in canonical form is

$$a_1 b_1 a_1^{-1} b_1^{-1} \cdots a_m b_m a_m^{-1} b_m^{-1}\, x_1 c_1 x_1^{-1} \cdots x_n c_n x_n^{-1} = 1$$

are $\chi = 2 - 2m - n$, $\beta = n$, $\omega = 0$.

Proof Let the numbers of vertices, edges and faces be V, E and F.

Then $V = n + 1$ (see Lemma 3.2), $E = 2m + 2n$ and $F = 1$, so

$$\chi = V - E + F = (n+1) - (2m+2n) + 1 = 2 - 2m - n.$$

The boundary number β is n, corresponding to the n holes $x_i c_i x_i^{-1}$.

Since no repeated edge appears twice in the same sense, the surface is orientable ($\omega = 0$). ■

It is useful to remember the formulas in the following way.

> ***Characteristic numbers for an orientable surface***
>
> For an orientable surface:
>
> - $\chi = 2 - 2$ (number of handles) $-$ (number of holes);
> - $\beta =$ (number of holes);
> - $\omega = 0$.

This result applies to the sphere too, which has no handles and no holes, giving $\chi = 2$, $\beta = 0$, $\omega = 0$.

Non-orientable surfaces

We saw at the start of this subsection that canonical form (c) corresponds to non-orientable surfaces. To find the corresponding characteristic numbers, we first need a preliminary result.

> ***Lemma 3.4***
>
> The method of inserting vertices applied to an edge equation of the form
>
> $$a_1 a_1 \cdots a_k a_k \, x_1 c_1 x_1^{-1} \cdots x_n c_n x_n^{-1} = 1 \quad (k > 0)$$
>
> yields
>
> $$Pa_1 Pa_1 \cdots Pa_k Pa_k \, Px_1 Q_1 c_1 Q_1 x_1^{-1} P \cdots Px_n Q_n c_n Q_n x_n^{-1} P = 1.$$

Proof Suppose that the edge a_1 starts at P. Then the process of inserting vertices into the edge equation begins

$$\underline{Pa_1 \, Pa_1} \, a_2 a_2 \cdots a_k a_k \, x_1 c_1 x_1^{-1} \cdots x_n c_n \, \underline{x_n^{-1} P} = 1,$$

showing that the edge a_1 also ends at P. It follows that the next edge a_2 starts at P, and therefore successively that all the edges a_i start and end at P. The result is

$$Pa_1 Pa_1 P \cdots Pa_k Pa_k P x_1 c_1 x_1^{-1} \cdots x_n c_n x_n^{-1} P = 1.$$

From this we see that x_1 starts at P, so the block $x_1 c_1 x_1^{-1}$ starts and ends at P. We now proceed exactly as in Lemma 3.2 to obtain

$$Pa_1 Pa_1 \cdots Pa_k Pa_k \, Px_1 Q_1 c_1 Q_1 x_1^{-1} P \cdots Px_n Q_n c_n Q_n x_n^{-1} P = 1.$$

There are $n + 1$ vertices: $P, Q_1, \ldots, Q_n$. ■

We deduce the following theorem.

Theorem 3.5

The characteristic numbers of any surface whose edge equation in canonical form is

$$a_1a_1 \cdots a_ka_k\, x_1c_1x_1^{-1} \cdots x_nc_nx_n^{-1} = 1 \quad (k > 0)$$

are $\chi = 2 - k - n$, $\beta = n, \omega = 1$.

Problem 3.2

Prove Theorem 3.5.

It is useful to remember the formulas in the following way.

Characteristic numbers for a non-orientable surface

For a non-orientable surface:

- $\chi = 2 -$ (number of cross-caps) $-$ (number of holes);
- $\beta =$ (number of holes);
- $\omega = 1$.

Uniqueness

We can now complete the proof of the Canonical Form Theorem, by showing that each surface has a unique canonical form.

Proof of uniqueness

The proof of existence showed that the edge equation of a surface can be reduced to $aa^{-1} = 1$ if the surface is a sphere, or to form (3.1). This in turn can be reduced to canonical form (b) if the surface is orientable, or to canonical form (c) if it is not. The only question that remains is whether the numbers m, n and k that specify forms (b) and (c) are determined uniquely by the original edge equation.

Recall from *Unit B2* that the characteristic numbers χ, β and ω of a surface are topological invariants.

For canonical form (b), we have $\chi = 2 - 2m - n$ and $\beta = n$, by Theorem 3.3. It follows that $n = \beta$ and $2m = 2 - \chi - \beta$, and so m and n are topological invariants. So the canonical form (b) is determined by the surface itself, and so is unique.

For canonical form (c), we have $\chi = 2 - k - n$ and $\beta = n$, by Theorem 3.5. It follows that $n = \beta$ and $k = 2 - \chi - \beta$, and so k and n are topological invariants. So the canonical form (c) is determined by the surface itself, and so is unique. ■

Problem 3.3

Show that the Euler characteristic of a surface cannot exceed 2.

3.2 Classifying surfaces

We now use the Canonical Form Theorem to prove the main theorem of this Block, one of the most important theorems of topology.

> ***Theorem 3.6 Classification Theorem for compact surfaces***
>
> Two compact surfaces are homeomorphic if and only if they have the same values for the characteristic numbers β, ω and χ.

Remark

We already know that the characteristic numbers are topological invariants. It is thus a necessary condition for two surfaces to be homeomorphic that they have the same characteristic numbers. The Classification Theorem shows that the equality of these three numbers is also sufficient.

Proof We first show that two orientable surfaces with the same characteristic numbers are homeomorphic.

We deduce from Theorem 3.3 that two possibilities exist for combinations of values for the characteristic numbers:

See Problem 3.3.

- $\chi = 2, \beta = 0, \omega = 0$
- $\chi < 2, \beta \geq 0, \omega = 0.$

In the first case, Theorem 3.3 tells us that the surface has no handles or holes, and so (from the existence part of Theorem 3.1) must have a canonical edge equation of the form $aa^{-1} = 1$. Therefore, if two surfaces have characteristic numbers $\chi = 2, \beta = 0, \omega = 0$, they must both be homeomorphic to the sphere, and hence to each other.

Now suppose the second case holds — that is, both surfaces have the same values for the characteristic numbers χ, β and ω, with $\chi < 2$ and $\beta \geq 0$. Both surfaces must have handles and/or holes (or else $\chi = 2$) and so must have canonical edge equations of form (b).

Note that $\omega = 0$, since we are considering orientable surfaces.

Suppose that the first surface has the canonical edge equation

$$a_1 b_1 a_1^{-1} b_1^{-1} \cdots a_m b_m a_m^{-1} b_m^{-1} x_1 c_1 x_1^{-1} \cdots x_n c_n x_n^{-1} = 1,$$

and that the second surface has the canonical edge equation

$$a_1 b_1 a_1^{-1} b_1^{-1} \cdots a_p b_p a_p^{-1} b_p^{-1} x_1 c_1 x_1^{-1} \cdots x_q c_q x_q^{-1} = 1.$$

Since both surfaces have the same value for β, we deduce from Theorem 3.3 that $n = q$. We also know that both surfaces have the same value for χ, so we deduce from Theorem 3.3 that

$$2 - 2m - n = 2 - 2p - q.$$

Since $n = q$, this means that $m = p$. It follows that the two surfaces are defined by the same canonical edge equation.

The passage from an edge equation to the surface it describes produces a surface that is unique up to homeomorphism, and so the two surfaces are homeomorphic.

The proof in the case of non-orientable surfaces is similar. We ask you to provide it in Problem 3.4. ■

Problem 3.4

Complete the proof of the Classification Theorem by showing that two *non-orientable* surfaces with the same characteristic numbers are homeomorphic.

3.3 Writing edge equations in canonical form

We now present a systematic method for reducing edge equations to canonical form. The method is similar to the steps in the proof of the existence of canonical forms in Theorem 3.1, in that we first assemble cross-caps, then handles, then holes. We illustrate the method by means of four worked problems.

Worked problem 3.1

Reduce the following edge equation to canonical form:

$$aba^{-1}cbdec^{-1}e = 1.$$

Find the characteristic numbers of the corresponding surface.

Solution

Step 1 Assemble cross-caps

We look systematically for repeated symbols appearing in the same sense (we check each a, each b, and so on).

First we find that the b symbols are repeated in the same sense. We say 'working with b' and use the Useful Moving Lemma to move to the right whatever lies between the two b symbols: this gives a string of the form $\cdots bb \cdots$. Then, using the Useful Moving Lemma again, we move across bb anything to the left of it, other than pairs of the form zz that may arise as we proceed.

We say 'working with b', rather than 'by the Useful Moving Lemma', because it specifies which symbols are being moved.

Thus, working with b, we obtain

$$aba^{-1}cbdec^{-1}e = 1 \;\rightarrow a\,\underline{bb}\,c^{-1}adec^{-1}e = 1 \;\rightarrow \underline{bb}\,ac^{-1}adec^{-1}e = 1.$$

Now we consider the part of the string that lies to the right of the block that we have dealt with (bb in this case), looking systematically for repeated symbols appearing the same sense. This time we find that the symbol a is repeated in this way, so we work with a:

$$bbac^{-1}adec^{-1}e = 1 \;\rightarrow bb\,\underline{aa}\,cdec^{-1}e = 1.$$

We continue the process, now considering the part of the string to the right of aa, looking systematically for repeated symbols appearing in the same sense. This time we find that the symbol e is repeated in this way, so we work with e:

$$bbaacdec^{-1}e = 1 \;\rightarrow bbaacd\,\underline{ee}\,c = 1 \;\rightarrow bbaa\,\underline{ee}\,cdc = 1.$$

Continuing the process, we now look to the right of ee and find that the symbol c is repeated in the same sense, so we work with c:

$$bbaaeecdc = 1 \;\rightarrow bbaaee\,\underline{cc}\,d^{-1} = 1.$$

The block to the right of cc contains no repeated symbols, so we have completed assembling the cross-caps.

Step 2 Assemble handles

In this case the block d^{-1} to the right of our assembly of cross-caps contains no strings of the form $\cdots y \cdots z \cdots y^{-1} \cdots z^{-1} \cdots$, so there are no handles.

Step 3 Assemble holes

Next we try to use the Jumping Lemma to assemble the holes. We first look to see if the block to the right of our assembly of cross-caps (and handles if there were any) contains pairs of symbols in opposite senses. In this case the block is d^{-1}, so there are none.

We now turn any unrepeated letters at the end of our assembly of cross-caps, handles and holes (in this case we only have cross-caps) into a hole. Here we make the isolated d^{-1} into a hole $fd^{-1}f^{-1}$ by adding a new symbol f to its left and its inverse f^{-1} to its right:

We saw that this is legitimate in the proof of Theorem 3.1.

$$bbaaeecc\,\underline{fd^{-1}f^{-1}} = 1.$$

If our equation now contains cross-caps and handles, we use the Replacement Lemma to replace the handles by cross-caps. In this case, we do not need to do this.

Thus the canonical form is

$$bbaaeeccfdf^{-1} = 1.$$

The canonical edge equation is of form (c).

We now use Theorem 3.5 to compute the characteristic numbers of the corresponding surface. The Euler characteristic is

$$\begin{aligned}\chi &= 2 - (\text{number of cross-caps}) - (\text{number of holes})\\ &= 2 - 4 - 1 = -3.\end{aligned}$$

There is one hole, so $\beta = 1$.

The surface is non-orientable: $\omega = 1$. ■

Worked problem 3.2

Reduce the following edge equation to canonical form:

$$ab^{-1}ca^{-1}c^{-1}debe^{-1} = 1.$$

Find the characteristic numbers of the corresponding surface.

Solution

Step 1 Assemble cross-caps

Since there are no symbols repeated in the same sense, there are no cross-caps to assemble.

Step 2 Assemble handles

We try to assemble the handles. We start by looking for the first pair of symbols occurring in the form $\cdots y \cdots z \cdots y^{-1} \cdots z^{-1} \cdots$. When we find them, we use the Assembling Lemma and 'working with y and z', we first push everything other than y, z, y^{-1}, and any assembly of cross-caps and handles at the start of the equation, to the right of z^{-1}.

Here we have a string of the form $\cdots a \cdots b^{-1} \cdots a^{-1} \cdots b \cdots$. Working with a and b^{-1}, and using the Assembling Lemma, we obtain

$$ab^{-1}ca^{-1}c^{-1}debe^{-1} = 1 \;\rightarrow\; \underline{ab^{-1}a^{-1}b}\,c^{-1}dece^{-1} = 1.$$

We now look for handles in the part of the string that lies to the right of the block $ab^{-1}a^{-1}b$. We see that the c^{-1} and e symbols occur in the form of a handle. So, working with c^{-1} and e, we have

$$ab^{-1}a^{-1}bc^{-1}dece^{-1} = 1 \rightarrow ab^{-1}a^{-1}b\,\underline{c^{-1}ece^{-1}}\,d = 1.$$

Step 3 Assemble holes

The block d to the right of our assembly of handles contains no more handles. Nor does it contain any repeated letters in opposite senses. It therefore remains to make the isolated d into a hole fdf^{-1}:

$$ab^{-1}a^{-1}bc^{-1}ece^{-1}\underline{fdf^{-1}} = 1.$$

This is the canonical form. The canonical edge equation is of form (b).

We now use Theorem 3.3 to compute the characteristic number of the corresponding surface. The Euler characteristic is

$$\begin{aligned}\chi &= 2 - 2\text{ (number of handles)} - \text{(number of holes)}\\ &= 2 - (2 \times 2) - 1 = -3,\end{aligned}$$

There is one hole, so $\beta = 1$.

The surface is orientable: $\omega = 0$. ■

Worked problem 3.3

Reduce the following edge equation to canonical form:

$$ab^{-1}cd^{-1}c^{-1}a^{-1}e = 1.$$

Find the characteristic numbers of the corresponding surface.

Solution

Step 1 Assemble cross-caps

There are no repeated symbols in the same sense, so there are no cross-caps to assemble.

Step 2 Assemble handles

There are no strings of the form $\cdots y \cdots z \cdots y^{-1} \cdots z^{-1} \cdots$, so there are no handles to assemble.

Step 3 Assemble holes

We now try to assemble the holes. We look for pairs of symbols occurring in opposite senses. We spot a, a^{-1} and c, c^{-1}. We consider the pair separated by the smallest number of symbols (in this case c and c^{-1}) and use the Jumping Lemma to give

$$ab^{-1}cd^{-1}c^{-1}a^{-1}e = 1 \rightarrow aba^{-1}cd^{-1}c^{-1}e = 1.$$

This one application of the Jumping Lemma has assembled both the hole cdc^{-1} and the hole $ab^{-1}a^{-1}$ at the start of the equation. There are no more pairs of symbols repeated in the same sense, so it remains to turn e into a hole fef^{-1} to give

$$ab^{-1}acdc^{-1}fef^{-1} = 1.$$

This canonical edge equation is of form (b). We now use Theorem 3.3 to compute the characteristic numbers. The Euler characteristic is

$$\begin{aligned}\chi &= 2 - 2\text{ (number of handles)} - \text{(number of holes)}\\ &= 2 - (2 \times 0) - 3 = -1.\end{aligned}$$

There are three holes, so $\beta = 3$.

The surface is orientable: $\omega = 0$. ■

In the following worked problem, we omit the explanations and show you a shorthand method for writing down the solution process. You should use this as a template for your solutions.

Worked problem 3.4

Reduce the following edge equation to canonical form:

$$abc^{-1}a^{-1}c^{-1}ded^{-1}e^{-1}b = 1.$$

Find the characteristic numbers of the corresponding surface.

Solution

$$abc^{-1}a^{-1}c^{-1}ded^{-1}e^{-1}b = 1$$

$$\begin{aligned}
&\to a\,\underline{bb}\,ede^{-1}d^{-1}cac = 1 && \text{(working with } b\text{)}\\
&\to \underline{bb}\,aede^{-1}d^{-1}cac = 1 && \text{(working with } b\text{)}\\
&\to bb\,\underline{aa}\,c^{-1}ded^{-1}e^{-1}c = 1 && \text{(working with } a\text{)}\\
&\to bbaa\,\underline{ded^{-1}e^{-1}}\,c^{-1}c = 1 && \text{(working with } d \text{ and } e\text{)}\\
&\to bbaa\,ded^{-1}e^{-1} = 1 && \text{(cancelling } c^{-1}c\text{)}\\
&\to bbaaddee = 1. && \text{(using the Replacement Lemma)}
\end{aligned}$$

This is the canonical form; it is of form (c).

The corresponding surface has Euler characteristic (Theorem 3.5.)

$$\begin{aligned}
\chi &= 2 - (\text{number of cross-caps}) - (\text{number of holes})\\
&= 2 - 4 - 0 = -2.
\end{aligned}$$

There are no holes, so $\beta = 0$.

The surface is non-orientable: $\omega = 1$. ■

We summarize the above methods in the following strategy.

> ***Strategy for obtaining the canonical form of an edge equation***
>
> 1. Assemble cross-caps aa (if any) using the Useful Moving Lemma and cycling if necessary.
> 2. Assemble handles $aba^{-1}b^{-1}$ (if any) using the Assembling Lemma.
> 3. Assemble holes xcx^{-1} using the Jumping Lemma and the following results:
>
>   - the edge equation
>
> $$a_1a_1 \cdots a_ka_kx_1cx_1^{-1} \cdots x_nc_nx_n^{-1}d = 1,$$
>
>   where d appears nowhere else in the equation, is equivalent to
>
> $$a_1a_1 \cdots a_ka_kx_1cx_1^{-1} \cdots x_nc_nx_n^{-1}ydy^{-1} = 1,$$
>
>   where y is a new symbol;
>   - the edge equation
>
> $$a_1b_1a_1^{-1}b_1^{-1} \cdots a_mb_ma_m^{-1}b_m^{-1}x_1cx_1^{-1} \cdots x_nc_nx_n^{-1}d = 1,$$
>
>   where d appears nowhere else in the equation, is equivalent to
>
> $$a_1b_1a_1^{-1}b_1^{-1} \cdots a_mb_ma_m^{-1}b_m^{-1}x_1cx_1^{-1} \cdots x_nc_nx_n^{-1}ydy^{-1} = 1,$$
>
>   where y is a new symbol.

Problem 3.5

Reduce each of the following edge equations to canonical form. In each case, find the characteristic numbers of the corresponding surface.

(a) $acbd^{-1}ab^{-1}cfdf^{-1} = 1$

(b) $bacdc^{-1}a^{-1}b^{-1}d^{-1} = 1$

(c) $bacdc^{-1}b^{-1}a^{-1}d^{-1} = 1$

3.4 Systems of edge equations

You may recall that in *Unit B2* we introduced the idea of a surface being represented by several polygons with edge identifications, each of which has an edge equation. You saw how such a system of edge equations can be combined to produce a single edge equation for the surface. Sometimes you may be asked to obtain the canonical form of the edge equation for a surface expressed as such a system of edge equations. All you need to do is first to combine the equations to give a single edge equation, and then to proceed as above, as illustrated in the following worked problem.

Unit B2, Subsection 3.3.

Worked problem 3.5

Reduce the following system of edge equations to canonical form, and find the characteristic numbers of the corresponding surface:

$$cac^{-1} = 1, dbd^{-1} = 1, abef^{-1} = 1.$$

Solution

$cac^{-1} = 1 \to a = c^{-1}c$ and $dbd^{-1} = 1 \to b = d^{-1}d$.

Substituting these into the remaining edge equation, we obtain

$$\begin{aligned} &c^{-1}cd^{-1}def^{-1} = 1 \\ &\to ef^{-1} = 1 \quad \text{(cancelling } c^{-1}c \text{ and } d^{-1}d\text{)} \\ &\to g = 1 \quad \text{(relabelling)} \\ &\to hgh^{-1} = 1 \quad \text{(creating a hole).} \end{aligned}$$

The canonical edge equation is of form (b).

The corresponding surface has Euler characteristic $\chi = 1$, boundary number $\beta = 1$, and is orientable ($\omega = 0$). ■

Problem 3.6

Reduce each of the following systems of edge equations to canonical form. In each case, find the characteristic numbers of the corresponding surface.

(a) $acb = 1$, $bcd^{-1} = 1$, $daf = 1$.

(b) $abca^{-1} = 1$, $ebdf^{-1} = 1$, $dcfg = 1$.

4 Connected sums of surfaces

After working through this section, you should be able to:

- explain what is meant by the *connected sum* of two compact surfaces;
- compute the characteristic numbers of a connected sum of compact surfaces;
- classify compact surfaces formed by the connected sum construction;
- compare the edge equation and connected sum descriptions of compact surfaces.

It is useful to be able to describe any compact surface in terms of its unique canonical form, but we still lack a convenient way of describing how the surface is constructed. The best description is provided by the *connected sum* of two or more surfaces. The connected sum construction is a way of gluing surfaces together to make new ones and it enables us to interpret the Classification Theorem geometrically. In this way we obtain a topological description of every compact surface, as follows:

- an orientable surface is either the sphere or a connected sum of toruses and/or closed discs;
- a non-orientable surface is a connected sum of projective planes and/or closed discs.

We use the following open-face letters for surfaces; for convenience, we also list their characteristic numbers.

$\mathbb{S} =$	sphere	$\chi = 2,\ \beta = 0,\ \omega = 0;$
$\mathbb{T} =$	torus	$\chi = 0,\ \beta = 0,\ \omega = 0;$
$\mathbb{D} =$	closed disc	$\chi = 1,\ \beta = 1,\ \omega = 0;$
$\mathbb{P} =$	projective plane	$\chi = 1,\ \beta = 0,\ \omega = 1;$
$\mathbb{K} =$	Klein bottle	$\chi = 0,\ \beta = 0,\ \omega = 1;$
$\mathbb{M} =$	Möbius band	$\chi = 0,\ \beta = 1,\ \omega = 1.$

4.1 The connected sum construction

The idea of a connected sum is as follows. We start with two compact surfaces S and T. We remove an open disc with boundary C from the surface S, and an open disc with boundary D from the surface T, thereby obtaining two new surfaces S' and T'. The new pieces of boundary, C on the surface S and D on the surface T, are homeomorphic to circles, and so are themselves homeomorphic. We let $f\colon C \to D$ be such a homeomorphism. To form the connected sum $S\#T$ of the surfaces S and T, we now identify points on the circles C and D that correspond under the homeomorphism f. The process is illustrated in Figure 4.1.

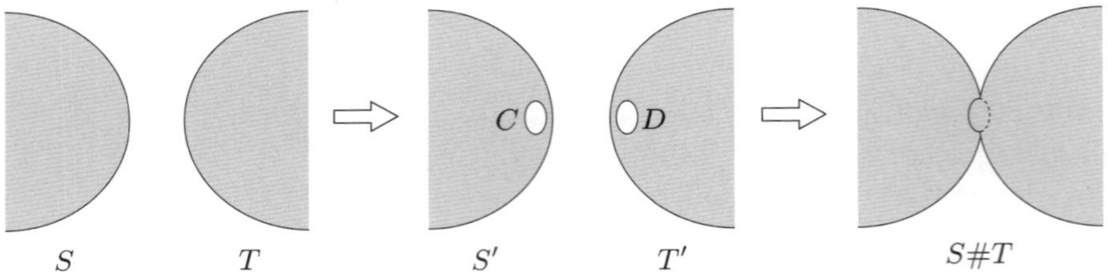

Figure 4.1

> ***Definition***
>
> Let S and T be two compact surfaces, and remove an open disc with boundary C from the surface S and an open disc with boundary D from the surface T. The **connected sum** $S\#T$ of S and T is the topological space obtained by identifying points on C and D that correspond under a homeomorphism $f\colon C \to D$.

We abbreviate the connected sum $S\#S$ of two copies of S to $2S$. We similarly write nS for the connected sum of n copies of S.

Before we make some remarks on this definition, we consider a few examples of connected sums.

$\mathbb{T}\#\mathbb{T}$

Figure 4.2 shows that the connected sum of two toruses is the 2-fold torus $2\mathbb{T}$.

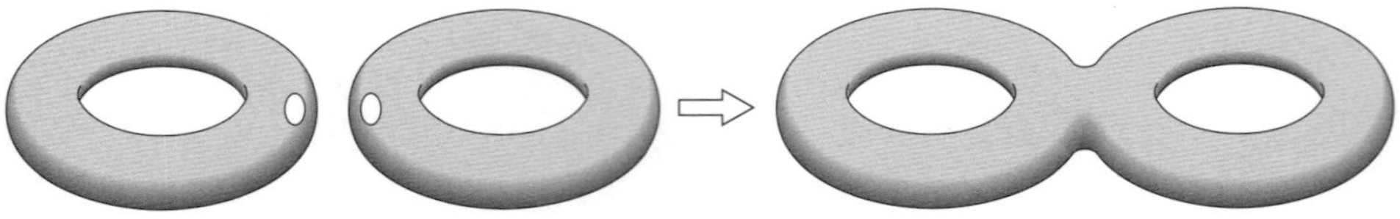

Figure 4.2

$S\#\mathbb{D}$

Figure 4.3 shows that, for any compact surface S, the connected sum $S\#\mathbb{D}$ is homeomorphic to the surface with an open disc removed (the surface with a hole added). We can see this by noting that the removal of an open disc from a closed disc produces an annulus that is homeomorphic to a cylinder. One end of the cylinder can be identified with the circle introduced on the surface S, while the other end is the boundary of the hole that must appear on S.

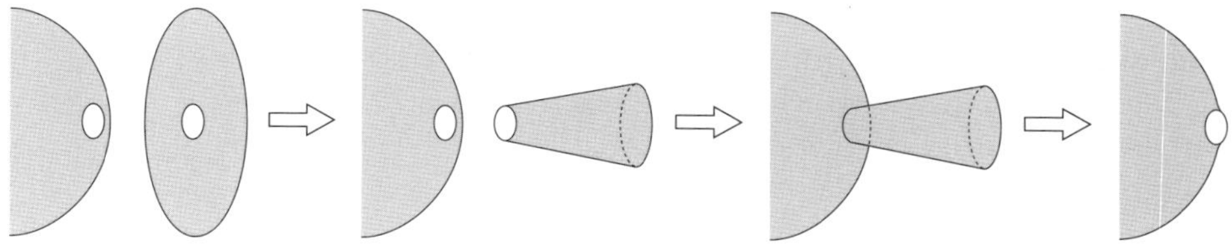

Figure 4.3

$S\#\mathbb{S}$

Figure 4.4 shows that, for any compact surface S, the connected sum $S\#\mathbb{S}$ is homeomorphic to the surface S. The sphere with an open disc removed contracts down to a closed disc (a cap) that spans the hole in the surface S.

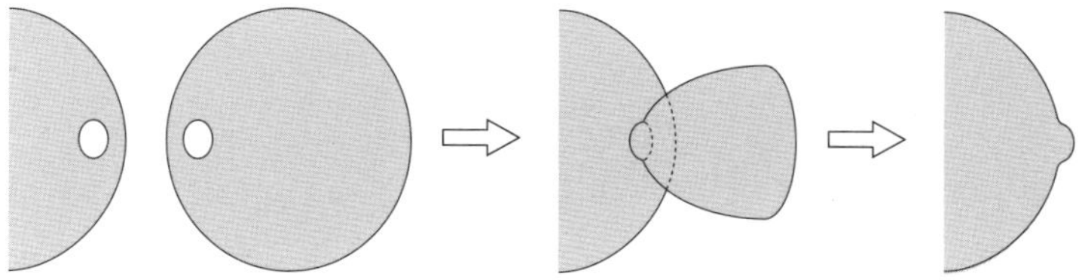

Figure 4.4

Remarks

(i) The connected sum $S\#T$ is always a surface.

(ii) The surface $S\#T$ is independent of the homeomorphism between the boundary circles C and D.

(iii) The surface $S\#T$ is independent of choice of open discs in each surface.

(iv) The connected sum operation is *commutative*: $S\#T = T\#S$.

(v) The connected sum operation is *associative*: $(S\#T)\#U = S\#(T\#U)$.

You may regard these results as being obvious, and indeed, essentially, they are. Nonetheless, we provide proofs of (ii) to (v) in the next subsection. A full proof of (i) can become rather technical, but essentially involves checking that the points along the 'join' do not violate the definition of a surface. We omit the details.

The definition of a surface is given at the end of Section 1 of *Unit B1*.

4.2 The characteristic numbers of a connected sum

The connected sum construction works for any pair of surfaces S and T, whether orientable or not, because we can always suppose that we have removed the discs from small patches on S and T. Let us now consider its effect on the characteristic numbers of the surfaces.

Suppose that we have two surfaces: S with Euler characteristic χ, boundary number β and orientability number ω, and T with Euler characteristic χ', boundary number β' and orientability number ω'. We want to find the characteristic numbers of their connected sum $S\#T$.

Euler characteristic

We remove an open disc from each surface, and we subdivide each surface so that each disc is a (polygonal) face of the surface to which it belongs. Without loss of generality, we can assume that each face has the same number of edges, n say. We can also assume that the connected sum construction identifies each of the n edges in S with a corresponding one of the n edges in T.

The effect on the Euler characteristic is as follows. Suppose that the subdivision of S has V vertices, E edges and F faces, and that the subdivision of T has V' vertices, E' edges and F' faces.

Removing the discs lowers the number of faces in each surface by 1, to $F - 1$ and $F' - 1$ respectively, so that after gluing, the subdivision of $S\#T$ has $F + F' - 2$ faces.

Identifying the boundaries of the discs identifies the n vertices on each boundary, thereby reducing the total number of vertices by n to $V + V' - n$.

It also identifies the n edges on each boundary, thereby decreasing the total number of edges by n to $E + E' - n$.

The Euler characteristic of $S\#T$ is therefore

$$(V + V' - n) - (E + E' - n) + (F + F' - 2)$$
$$= (V - E + F) + (V' - E' + F') - 2 = \chi + \chi' - 2.$$

Boundary number

Removing an open disc from each surface raises the boundary number by 1 on each surface, giving a total increase of 2. But identifying the edges of the removed discs removes these two new boundary curves. The net effect is that no boundary components are introduced and none are removed, so the boundary number of $S\#T$ is $\beta + \beta'$.

Orientability number

When a surface contains a Möbius band, we can move the Möbius band away from a given open disc in the surface, and can thus remove the disc from the surface without affecting the Möbius band. So the connected sum of two surfaces, at least one of which is non-orientable, is non-orientable.

If both surfaces $S\#T$ are orientable, then the connected sum $S\#T$ is orientable. To see this, suppose that each point of S is the centre of a small clock which is oriented clockwise. Because S is orientable, the orientations are compatible. Now do the same for the surface T. In the surface S, pick the disc to be removed and observe that there is a clock at each point of the boundary of this disc, and all these clocks have a coherent sense of clockwise. Again, do the same for the surface T.

We make no claim about the compatibility of clocks on different surfaces.

After removing the disc you still have half a clock!

Now glue the surfaces S and T along the boundaries of the discs, and look at the 'half-clocks' on each side of the identified boundary. There are only two possibilities: either they fit together at every point to form a proper clock, or *at every point* they have opposite orientations. In the former case the connected sum has a compatible set of clocks everywhere and is orientable. In the latter case, reverse the orientation of every clock on the surface T. Then the connected sum has a compatible set of clocks everywhere, and is orientable.

It follows that the connected sum of two surfaces is orientable if and only if both the surfaces are orientable.

We have thus established the following result.

> ***Theorem 4.1***
>
> The connected sum $S\#T$ of two compact surfaces S and T has the following characteristic numbers:
>
> $\chi(S\#T) = \chi(S) + \chi(T) - 2;$
>
> $\beta(S\#T) = \beta(S) + \beta(T);$
>
> $\omega(S\#T) = 0$ if and only if $\omega(S) = \omega(T) = 0.$

$\omega(S\#T) = 1$ if and only if $\omega(S)$ and $\omega(T)$ are not both 0.

Remark

Theorem 4.1, together with the Classification Theorem, tells us that the connected sum $S\#T$ of any two compact surfaces S and T is unique up to homeomorphism. Notice also that at no stage did the computations for the characteristic numbers of $S\#T$ depend on the choice of open discs or the choice of homeomorphism between the boundary circles. We have thus proved Remarks (ii) and (iii) from Subsection 4.1. Theorem 4.1 also allows us to prove Remarks (iv) and (v), as the following problem asks you to demonstrate.

Problem 4.1

Let S, T and U be compact surfaces.

(a) Show that the characteristic numbers of $S\#T$ and $T\#S$ are the same, and deduce that these surfaces are homeomorphic.

Commutativity.

(b) Show that the characteristic numbers of $(S\#T)\#U$ and $S\#(T\#U)$ are the same, and deduce that these surfaces are homeomorphic.

Associativity.

Problem 4.2

(a) Write down the characteristic numbers of the surfaces

$$S_1 = 2\mathbb{T}\#3\mathbb{D},\ S_2 = 3\mathbb{T},\ S_3 = 3\mathbb{P}\#4\mathbb{D}.$$

(b) Write down the characteristic numbers of the six connected sums

$$S_1\#S_1,\ S_1\#S_2,\ S_1\#S_3,\ S_2\#S_2,\ S_2\#S_3,\ S_3\#S_3.$$

4.3 Connected sums and edge equations

We now investigate how to relate edge equations in canonical form to connected sums.

We begin with orientable surfaces. We know from Theorem 3.1 that every orientable surface other than the sphere has a canonical edge equation of the form

$$a_1b_1a_1^{-1}b_1^{-1}\cdots a_mb_ma_m^{-1}b_m^{-1}\,x_1c_1x^{-1}\cdots x_nc_nx_n^{-1} = 1.$$

We know that the blocks $a_ib_ia_i^{-1}b_i^{-1}$ represent handles and that the blocks $x_jc_jx_j^{-1}$ represent holes. We have also seen that adding a handle to a surface is equivalent to adding a torus and that adding a hole is equivalent to adding a closed disc. So we suspect that the above edge equation represents the connected sum $m\mathbb{T}\#n\mathbb{D}$.

To prove this, we use the Classification Theorem. We know from Theorem 3.3 that the surface with the above edge equation has characteristic numbers $\chi = 2 - 2m - n$, $\beta = n$ and $\omega = 0$. We know from Theorem 4.1 that the connected sum $m\mathbb{T}\#n\mathbb{D}$ has Euler characteristic

$$\begin{aligned}\chi(m\mathbb{T}\#n\mathbb{D}) &= \chi(m\mathbb{T}) + \chi(n\mathbb{D}) - 2 \\ &= -2(m-1) + (n - 2(n-1)) - 2 \\ &= 2 - 2m - n.\end{aligned}$$

Also, $\beta = n$ and $\omega = 0$. Therefore, by the Classification Theorem, the edge equation and the connected sum represent the same surface.

Theorem 4.2

The edge equation

$$a_1 b_1 a_1^{-1} b_1^{-1} \cdots a_m b_m a_m^{-1} b_m^{-1} x_1 c_1 x^{-1} \cdots x_n c_n x_n^{-1} = 1 \quad (m \geq 0, n \geq 0)$$

is the canonical edge equation for the orientable surface $m\mathbb{T}\#n\mathbb{D}$.

For a non-orientable surface, Theorem 3.1 tells that the canonical edge equation is of the form

$$a_1 a_1 \cdots a_k a_k x_1 c_1 x_1^{-1} \cdots x_n c_n x_n^{-1} = 1.$$

We know that each block $a_i a_i$ represents a cross-cap — that is, a projective plane — and that each block $x_j c_j x_j^{-1}$ represents a hole. So this time the edge equation represents the connected sum $k\mathbb{P}\#n\mathbb{D}$.

Theorem 4.3

The edge equation

$$a_1 a_1 \cdots a_k a_k\, x_1 c_1 x_1^{-1} \cdots x_n c_n x_n^{-1} = 1 \quad (k > 0, n \geq 0)$$

is the canonical edge equation for the non-orientable surface $k\mathbb{P}\#n\mathbb{D}$.

Problem 4.3

Prove Theorem 4.3.

Theorems 4.2 and 4.3 can now be used together with the Canonical Form Theorem and the Classification Theorem to deduce the following result.

Theorem 4.4 Classification of compact surfaces

A compact surface is one of the following:

- the sphere $\mathbb{S}$;
- an orientable surface of the form $m\mathbb{T}\#n\mathbb{D}$ $(m \geq 0, n \geq 0)$;
- a non-orientable surface of the form $k\mathbb{P}\#n\mathbb{D}$ $(k > 0, n \geq 0)$.

Remarks

(i) The surface $m\mathbb{T}\#n\mathbb{D}$ is an m-fold torus with n holes. Since, for any surface S, $\mathbb{S}\#S$ is homeomorphic to S, we can also think of $m\mathbb{T}\#n\mathbb{D}$ as a sphere with m handles and n holes. If $m = 0$, we have a sphere with n holes.

(ii) Similarly, the surface $k\mathbb{P}\#n\mathbb{D}$ can be thought of as a sphere with k cross-caps and n holes.

An important special case of Theorem 4.4 concerns compact surfaces without boundary. In this case there are no holes, so $n = 0$, and we obtain the following result.

Theorem 4.5

A compact surface without boundary is one of the following:

- the sphere $\mathbb{S}$;
- an orientable surface of the form $m\mathbb{T}$ — that is, a sphere with m handles;
- a non-orientable surface of the form $k\mathbb{P}$ — that is, a sphere with k cross-caps.

This result will be used in *Unit B4*.

It is clear that Theorem 4.5 takes account of several of our standard compact surfaces without boundary: the sphere $\mathbb{S}$; the torus $\mathbb{T}$; the m-fold torus $m\mathbb{T}$; the projective plane $\mathbb{P}$. But what about the Klein bottle $\mathbb{K}$? As it is non-orientable, we can deduce that it is of the form $k\mathbb{P}$ for some k.

Problem 4.4

Show that the Klein bottle is the connected sum of two copies of the projective plane — that is, $\mathbb{K}$ is homeomorphic to $2\mathbb{P}$.

Thus the Klein bottle is a sphere with two cross-caps.

4.4 Connected sums of connected sums

Suppose we want to form the connected sum of two surfaces given in connected sum form.

We know already that if one surface is the sphere $\mathbb{S}$, then the connected sum is simply the other surface.

What if both are orientable, but neither is the sphere? You are unlikely to be surprised by the following result.

Theorem 4.6

The connected sum of $m\mathbb{T}\#n\mathbb{D}$ and $m'\mathbb{T}\#n'\mathbb{D}$ is

$$(m + m')\mathbb{T}\#(n + n')\mathbb{D}.$$

Proof Using the commutative and associative laws for the connected sum operation, we have

$$\begin{aligned}(m\mathbb{T}\#n\mathbb{D})\#(m'\mathbb{T}\#n'\mathbb{D}) &= (m\mathbb{T}\#m'\mathbb{T})\#(n\mathbb{D}\#n'\mathbb{D})\\ &= (m+m')\mathbb{T}\#(n+n')\mathbb{D}.\end{aligned}$$ ■

We could also prove this using characteristic numbers.

When both surfaces are non-orientable, we have the following equally unsurprising result.

Theorem 4.7

The connected sum of $m\mathbb{P}\#n\mathbb{D}$ and $m'\mathbb{P}\#n'\mathbb{D}$ is

$$(m+m')\mathbb{P}\#(n+n')\mathbb{D}.$$

The proof, which we omit, is very similar to that of Theorem 4.6.

However, what happens if one surface is orientable and the other is not? In other words, what happens if one surface has the form $m\mathbb{T}\#n\mathbb{D}$ and the other has the form $k\mathbb{P}\#n'\mathbb{D}$? Applying the commutative and associative laws to the connected sum gives

$$(m\mathbb{T}\#n\mathbb{D})\#(k\mathbb{P}\#n'\mathbb{D}) = k\mathbb{P}\#m\mathbb{T}\#(n+n')\mathbb{D}.$$

The result is not in one of the standard forms of Theorem 4.4.

To put it in standard form, we recall the proof of the Canonical Form Theorem, where we used the Replacement Lemma to transform an edge equation consisting of cross-caps, handles and holes into one involving simply cross-caps and holes. That would suggest that here we can replace $k\mathbb{P}\#m\mathbb{T}\#(n+n')\mathbb{D}$ by $(k+2m)\mathbb{P}\#(n+n')\mathbb{D}$.

Theorem 4.8

The connected sum of $m\mathbb{T}\#n\mathbb{D}$ and $k\mathbb{P}\#n'\mathbb{D}$ is

$$(2m+k)\mathbb{P}\#(n+n')\mathbb{D}.$$

Problem 4.5

Show that $\mathbb{T}\#\mathbb{P}$ is homeomorphic to $3\mathbb{P}$.

Problem 4.6

Prove Theorem 4.8 by calculating the characteristic numbers of $S = m\mathbb{T}\#n\mathbb{D}$, $T = k\mathbb{P}\#n'\mathbb{D}$, $S\#T$ and $U = (2m+k)\mathbb{P}\#(n+n')\mathbb{D}$.

4.5 Classifying surfaces by characteristic numbers

The above theorems allow us to classify compact surfaces as connected sums in terms of their Euler characteristic and boundary number.

The first step is to recall the corollary to the Classification Theorem that the maximum Euler characteristic of a compact surface is 2, which is the Euler characteristic of the sphere. We also note, from Theorem 4.4, that all compact surfaces other than the sphere can be obtained from the sphere by adding handles (toruses) and holes (closed discs), or by adding cross-caps (projective planes) and holes (closed discs). Theorem 4.1 tells us that:

Problem 3.3.

Remarks (i) and (ii).

- adding a torus lowers the Euler characteristic by 2, and leaves the boundary number unaltered;
- adding a disc lowers the Euler characteristic by 1, and raises the boundary number by 1;
- adding a projective plane lowers the Euler characteristic by 1, and leaves the boundary number unaltered.

We can therefore build up a table of compact surfaces, starting with the sphere. The rows of the table correspond to the value of χ and the columns to the value of β. We generate the $(k+1)$th row as follows: to the surfaces in the kth row we add either a closed disc or a projective plane, and to the surfaces in the $(k-1)$th row we add a torus. We then use the homeomorphism between $\mathbb{T}\#\mathbb{P}$ and $3\mathbb{P}$ to eliminate any repetition. The first few rows of the table are as follows.

Problem 4.5.

	$\beta = 0$	$\beta = 1$	$\beta = 2$	$\beta = 3$
$\chi = 2$	$\mathbb{S}$			
$\chi = 1$	$\mathbb{P}$	$\mathbb{D}$		
$\chi = 0$	$\mathbb{T}, 2\mathbb{P}$	$\mathbb{P}\#\mathbb{D}$	$2\mathbb{D}$	
$\chi = -1$	$3\mathbb{P}$	$\mathbb{T}\#\mathbb{D}, 2\mathbb{P}\#\mathbb{D}$	$\mathbb{P}\#2\mathbb{D}$	$3\mathbb{D}$

Problem 4.7

Construct the next row of the table, for $\chi = -2$.

By attempting Problem 4.7, you will appreciate that there is much repetition in applying the above strategy for constructing the table. Fortunately, we can prove two results that considerably reduce the amount of work required at each stage. For orientable surfaces, the result we need is as follows.

Theorem 4.9

There is an orientable surface with Euler characteristic $\chi \leq 2$ and boundary number $\beta \geq 0$ if and only if

$$\chi + \beta \text{ is even} \quad \text{and} \quad \chi + \beta \leq 2.$$

Under these conditions, the surface is the sphere if $\chi = 2$, and otherwise it is

$$(1 - \tfrac{1}{2}(\chi + \beta))\mathbb{T}\#\beta\mathbb{D}.$$

Proof If $\chi = 2$, we already know that there exists just one (orientable) compact surface — namely, the sphere.

If $\chi < 2$, Theorem 4.4 tells us that any orientable surface must have the form $\alpha\mathbb{T}\#\beta\mathbb{D}$, for integers $\alpha \geq 0$, $\beta \geq 0$. By Theorem 4.1, this surface has Euler characteristic $\chi = 2 - 2\alpha - \beta$ and boundary number β. The equation $\chi = 2 - 2\alpha - \beta$ gives $\alpha = 1 - \frac{1}{2}(\chi + \beta)$, which is a non-negative integer if and only if $\chi + \beta$ is even and $\chi + \beta \leq 2$. ■

Remark

The condition $\chi + \beta \leq 2$ provides another proof that all orientable surfaces have characteristic number $\chi \leq 2$.

Theorem 4.9 enables us to find the orientable surfaces with given Euler characteristic $\chi < 2$. There is exactly one surface for each value of β for which $\chi + \beta$ is even (so χ and β have the same parity) and for which $\beta \leq 2 - \chi$.

For non-orientable surfaces, we have the following result.

Theorem 4.10

There is a non-orientable surface with Euler characteristic $\chi < 2$ and boundary number $\beta \geq 0$ if and only if

$$\chi + \beta \leq 1.$$

Under these conditions, this surface is

$$(2 - (\chi + \beta))\mathbb{P}\#\beta\mathbb{D}.$$

Proof Theorem 4.4 tells us that any non-orientable surface must have the form $\alpha\mathbb{P}\#\beta\mathbb{D}$, for integers $\alpha > 0$, $\beta \geq 0$. By Theorem 4.1, this surface has Euler characteristic $\chi = 2 - \alpha - \beta$ and boundary number β. The equation $\chi = 2 - \alpha - \beta$ gives $\alpha = 2 - (\chi + \beta)$, which is a positive integer if and only if $\chi + \beta \leq 1$. ■

Remark

The condition $\chi + \beta \leq 1$ provides another proof that all non-orientable surfaces have characteristic number $\chi < 2$.

Theorem 4.10 enables us to find the non-orientable surfaces with a given Euler characteristic $\chi < 2$. There is exactly one surface for each value of β for which $\beta \leq 1 - \chi$.

You can now use Theorems 4.9 and 4.10 to verify the entries in the above table and your answer to Problem 4.7.

Problem 4.8

(a) Find all the orientable surfaces with Euler characteristic -3.

(b) Find all the non-orientable surfaces with Euler characteristic -3.

Solutions to problems

1.1 We obtain the two polygons

$ac^{-1}ddf^{-1}y^{-1} = 1 \quad \text{and} \quad yf^{-1}ab^{-1} = 1.$

To glue these polygons along the edge a, we rewrite the first edge equation as $a(c^{-1}ddf^{-1}y^{-1}) = 1$, leading to

$a = (c^{-1}ddf^{-1}y^{-1})^{-1} = yfd^{-1}d^{-1}c.$

We now substitute this expression for a into the second edge equation, giving

$yf^{-1}yfd^{-1}d^{-1}cb^{-1} = 1.$

(Note that there are several different ways of writing this edge equation.)

1.2

(a) $\boldsymbol{A}^{-1} = (aba^{-1}cdc^{-1}b)^{-1} = b^{-1}cd^{-1}c^{-1}ab^{-1}a^{-1}$

(b) $\boldsymbol{A}^{-1} = (bcab^{-1}c^{-1}a)^{-1} = a^{-1}cba^{-1}c^{-1}b^{-1}$

1.3 The cut-and-glue operation is defined by $y = a\boldsymbol{C}$, where $\boldsymbol{C} = c^{-1}ddf^{-1}$.

1.4 **(a)** Consider the cut-and-glue operation defined by $y = x\boldsymbol{B}$, so that $x = y\boldsymbol{B}^{-1}$. This operation transforms the edge equation $\boldsymbol{A}x\boldsymbol{B}\boldsymbol{C}x\boldsymbol{D} = 1$ into $\boldsymbol{A}y\boldsymbol{C}y\boldsymbol{B}^{-1}\boldsymbol{D} = 1$. Thus, the given equations are equivalent.

(b) $\boldsymbol{A}\boldsymbol{B}\boldsymbol{C} = 1$ is equivalent to $\boldsymbol{A}\boldsymbol{B}yy^{-1}\boldsymbol{C} = 1$, for any y. Putting $y = x^{-1}$ gives the result.

2.1 **(a)** Let $\boldsymbol{A}$ and $\boldsymbol{C}$ be the empty block, and let $\boldsymbol{B}$ and $\boldsymbol{D}$ be the block consisting only of y. Then, by the Moving Lemma, part (a), $\boldsymbol{A}x\boldsymbol{B}\boldsymbol{C}x\boldsymbol{D} = xyxy = 1$ is equivalent to $\boldsymbol{A}x\boldsymbol{C}x\boldsymbol{B}^{-1}\boldsymbol{D} = xxy^{-1}y = xx = 1$. (Recall that we can drop expressions of the form $y^{-1}y$ from edge equations.)

(b) Let $\boldsymbol{A}$, $\boldsymbol{C}$ and $\boldsymbol{D}$ be the empty block and let $\boldsymbol{B}$ be the block $xyzy^{-1}$. Then, by the Moving Lemma, part (a), we have $\boldsymbol{A}w\boldsymbol{B}\boldsymbol{C}w\boldsymbol{D} = w(xyzy^{-1})w = 1$ is equivalent to

$\boldsymbol{A}w\boldsymbol{C}w\boldsymbol{B}^{-1}\boldsymbol{D} = ww(yz^{-1}y^{-1}x^{-1}) = 1.$

(We could have proved both parts by using part (b) of the Moving Lemma instead.)

2.2 **(a)** $xx\boldsymbol{E}\boldsymbol{D} = 1 \to \boldsymbol{D}xx\boldsymbol{E} = 1$ (cycling)

$\to xx\boldsymbol{D}\boldsymbol{E} = 1$

(using UML, with $\boldsymbol{B} = \boldsymbol{D}, \boldsymbol{C} = \boldsymbol{E}$ and $\boldsymbol{A}$ empty).

(b) $auvuxxab = 1 \to axxuvuab = 1$

(using UML to move uvu across xx).

2.3 Let $\boldsymbol{A} = a$, $\boldsymbol{B} = b$, $\boldsymbol{C} = c$ and $\boldsymbol{D} = b^{-1}d$.

Inserting vertices in the edge equation $\boldsymbol{A}\boldsymbol{B}x\boldsymbol{C}x\boldsymbol{D} = abxcxb^{-1}d = 1$, we obtain

$PaQbRxRcRxRb^{-1}QdP = 1,$

so there are 3 vertices and 5 edges. The Euler characteristic of the corresponding surface is $3 - 5 + 1 = -1$.

Inserting vertices in the edge equation $\boldsymbol{A}x\boldsymbol{C}x\boldsymbol{B}\boldsymbol{D} = axcxbb^{-1}d = axcxd = 1$, we obtain

$PaQxRcQxRdP = 1,$

so there are 3 vertices and 4 edges. The Euler characteristic of the corresponding surface is $3 - 4 + 1 = 0$.

So the two edge equations are not equivalent.

2.4 $axxuvuab = 1 \to xxuvuaba = 1$ (cycling)

$\to xxuvuaab^{-1} = 1$

(using UML to bring the a terms together)

$\to xxaauvub^{-1} = 1$

(using UML to move uvu across aa)

$\to xxaauuv^{-1}b^{-1} = 1$

(using UML to bring the u terms together).

2.5 Apply the Jumping Lemma with $\boldsymbol{A}$ replaced by $\boldsymbol{D}$, $\boldsymbol{B}$ replaced by $u\boldsymbol{B}u^{-1}$ and $\boldsymbol{D}$ replaced by $\boldsymbol{E}$; then $u\boldsymbol{B}u^{-1}$ jumps over $x\boldsymbol{C}x^{-1}$, as required.

2.6 $\boldsymbol{A}xyx^{-1}y^{-1}\boldsymbol{B} = 1$

$\to \boldsymbol{A}\boldsymbol{B}xyx^{-1}y^{-1} = 1$

(AL with $\boldsymbol{C}$, $\boldsymbol{D}$, $\boldsymbol{E}$, $\boldsymbol{F}$ empty)

$\to xyx^{-1}y^{-1}\boldsymbol{A}\boldsymbol{B} = 1$ (cycling).

2.7 **(a)** $abcacb^{-1} = 1$

$\to aac^{-1}b^{-1}cb^{-1} = 1$ (UML)

$\to aac^{-1}c^{-1}b^{-1}b^{-1} = 1$ (UML)

$\to xxyyzz = 1$ (relabelling).

(b) $aba^{-1}b^{-1}e = 1 \to aba^{-1}b^{-1}ed^{-1}d = 1$ (adding $d^{-1}d$)

$\to daba^{-1}b^{-1}ed^{-1} = 1$ (cycling)

$\to aba^{-1}b^{-1}ded^{-1} = 1$ (AL).

2.8 **(a)** Inserting vertices in the edge equation $xxaba^{-1}b^{-1} = 1$, we obtain

$PxPxPaPbPa^{-1}Pb^{-1}P = 1.$

So the Euler characteristic of the corresponding surface is $1 - 3 + 1 = -1$. A surface with edge equation $xx = 1$ is a projective plane and so has Euler characteristic 1. So the edge equations are not equivalent.

(b) Set $\boldsymbol{C} = xxa$, $\boldsymbol{A} = b$, $\boldsymbol{B} = a^{-1}b^{-1}$. The edge equation $\boldsymbol{C}\boldsymbol{A}\boldsymbol{B} = 1$ becomes $xxaba^{-1}b^{-1} = 1$. The edge equation $\boldsymbol{C}\boldsymbol{B}\boldsymbol{A} = 1$ becomes $xxaa^{-1}b^{-1}b = 1$, which (by removing spheres) is equivalent to $xx = 1$. By (a), $xxaba^{-1}b^{-1} = 1$ and $xx = 1$ are not equivalent. It follows that an edge equation of the form $\boldsymbol{C}\boldsymbol{A}\boldsymbol{B} = 1$ is not necessarily equivalent to one of the form $\boldsymbol{C}\boldsymbol{B}\boldsymbol{A} = 1$.

3.1 (a) torus: form (b) with $m = 1$, $n = 0$;
2-fold torus: form (b) with $m = 2$, $n = 0$;
projective plane: by the result of Problem 2.1(a) this is equivalent to $aa = 1$, and is thus of form (c) with $k = 1$, $n = 0$.

(b) By the Useful Moving Lemma with $\boldsymbol{A}$ empty, the edge equations $x\boldsymbol{B}^{-1}x\boldsymbol{C} = 1$ and $xx\boldsymbol{BC} = 1$ are equivalent. Writing $x = a$, $\boldsymbol{B}^{-1} = b$, $\boldsymbol{C} = b^{-1}$, we deduce that the edge equations $abab^{-1} = 1$ and $aab^{-1}b^{-1}$ are equivalent. The Klein bottle therefore has an edge equation of form (c) with $k = 2$, $n = 0$.

3.2 Let the numbers of vertices, edges and faces be V, E and F.

Then $V = n + 1$, $E = k + 2n$ and $F = 1$, so

$$\begin{aligned}\chi &= V - E + F = (n+1) - (k+2n) + 1\\ &= 2 - k - n.\end{aligned}$$

The boundary number β is n, corresponding to the n holes $x_i c_i x_i^{-1}$.

Since $k > 0$, at least one repeated edge appears twice in the same sense, so the surface is non-orientable ($\omega = 1$).

3.3 If the surface is orientable, its Euler characteristic is

2 − 2 (number of handles) − (number of holes).

Since the numbers of handles and holes are non-negative, the Euler characteristic is at most 2.

If the surface is non-orientable, its Euler characteristic is

2 − (number of cross-caps) − (number of holes).

Since the numbers of cross-caps and holes are non-negative, the Euler characteristic is at most 2.

3.4 Suppose we have two non-orientable surfaces with the same values for the characteristic numbers χ, β and ω. (We must have $\omega = 1$.) The existence part of Theorem 3.1 tells us that the surfaces must have canonical edge equations of form (c). Suppose that the first surface has the canonical edge equation

$$a_1a_1 \cdots a_ka_k\, x_1c_1x_1^{-1} \cdots x_nc_nx_n^{-1} = 1,$$

and that the second surface has the canonical edge equation

$$a_1a_1 \cdots a_pa_p\, x_1c_1x_1^{-1} \cdots x_qc_qx_q^{-1} = 1.$$

Since both surfaces have the same value for β, we deduce from Theorem 3.5 that $n = q$. We also know that both surfaces have the same value for χ, so we deduce from Theorem 3.5 that

$$2 - k - n = 2 - p - q.$$

Since $n = q$, this means that $k = p$. It follows that the two surfaces are defined by the same canonical edge equation. The passage from an edge equation to the surface it describes produces a surface that is unique up to homeomorphism, and so the two surfaces are homeomorphic.

3.5 (a) $acbd^{-1}ab^{-1}cfdf^{-1} = 1$

$\to aadb^{-1}c^{-1}b^{-1}cfdf^{-1} = 1$ (working with a)
$\to aaddf^{-1}c^{-1}bcbf^{-1} = 1$ (working with d)
$\to aaddf^{-1}f^{-1}b^{-1}c^{-1}b^{-1}c = 1$ (working with f^{-1})
$\to aaddf^{-1}f^{-1}b^{-1}b^{-1}cc = 1$ (working with b^{-1}).

The canonical edge equation is of form (c).

The corresponding surface has Euler characteristic

$$\begin{aligned}\chi &= 2 - (\text{number of cross-caps}) - (\text{number of holes})\\ &= 2 - 5 - 0 = -3.\end{aligned}$$

It has boundary number $\beta = 0$, and is non-orientable ($\omega = 1$).

(b) $bacdc^{-1}a^{-1}b^{-1}d^{-1} = 1$

$\to cdc^{-1}d^{-1}baa^{-1}b^{-1} = 1$ (working with c and d)
$\to cdc^{-1}d^{-1}bb^{-1} = 1$ (cancelling aa^{-1})
$\to cdc^{-1}d^{-1} = 1$ (cancelling bb^{-1}).

The canonical edge equation is of form (b).

The corresponding surface has Euler characteristic

$$\begin{aligned}\chi &= 2 - 2(\text{number of handles}) - (\text{number of holes})\\ &= 2 - 2 - 0 = 0.\end{aligned}$$

It has boundary number $\beta = 0$, and is orientable ($\omega = 0$).

(c) $bacdc^{-1}b^{-1}a^{-1}d^{-1} = 1$

$\to bab^{-1}a^{-1}cdc^{-1}d^{-1} = 1$ (working with b and a).

The canonical edge equation is of form (b).

The corresponding surface has Euler characteristic

$$\begin{aligned}\chi &= 2 - 2(\text{number of handles}) - (\text{number of holes})\\ &= 2 - 4 - 0 = -2.\end{aligned}$$

It has boundary number $\beta = 0$, and it is orientable ($\omega = 0$).

3.6 (a) $bcd^{-1} = 1 \to b = dc^{-1}$.

Substituting this in the first edge equation gives

$$acdc^{-1} = 1.$$

Also,

$$daf = 1 \to d = f^{-1}a^{-1}.$$

Substituting this into the edge equation above gives $acf^{-1}a^{-1}c^{-1} = 1$

$\to aca^{-1}c^{-1}f^{-1} = 1$ (working with a and c)
$\to aca^{-1}c^{-1}gf^{-1}g^{-1} = 1$ (creating a hole.)

The canonical edge equation is of form (b).

The corresponding surface has Euler characteristic $\chi = -1$, boundary number $\beta = 1$, and is orientable ($\omega = 0$).

(b) $ebdf^{-1} = 1 \to b = e^{-1}fd^{-1}$

$dcfg = 1 \to c = d^{-1}g^{-1}f^{-1}$

Substituting these expressions into the first edge equation, gives

$ae^{-1}fd^{-1}d^{-1}g^{-1}f^{-1}a^{-1} = 1$
$\to d^{-1}d^{-1}ae^{-1}fg^{-1}f^{-1}a^{-1} = 1$ (working with d^{-1})
$\to d^{-1}d^{-1}fg^{-1}f^{-1}ae^{-1}a^{-1} = 1$ (Jumping Lemma)

The canonical edge equation is of form (c).

The corresponding surface has Euler characteristic $\chi = -1$, boundary number $\beta = 2$, and is non-orientable ($\omega = 1$).

4.1 **(a)** By Theorem 4.1, we have:

$$\chi(S\#T) = \chi(S) + \chi(T) - 2$$
$$= \chi(T) + \chi(S) - 2 = \chi(T\#S);$$
$$\beta(S\#T) = \beta(S) + \beta(T)$$
$$= \beta(T) + \beta(S) = \beta(T\#S);$$
$$\omega(S\#T) = 0 \iff \omega(S) = \omega(T) = 0$$
$$\iff \omega(T\#S) = 0,$$

so $\omega(S\#T) = \omega(T\#S)$.

So the characteristic numbers of the surfaces $S\#T$ and $T\#S$ are the same. By the Classification Theorem, these surfaces are homeomorphic.

(b) By Theorem 4.1, we have:

$$\chi((S\#T)\#U) = \chi(S\#T) + \chi(U) - 2$$
$$= \chi(S) + \chi(T) + \chi(U) - 4$$
$$= \chi(S) + \chi(T\#U) - 2$$
$$= \chi(S\#(T\#U));$$
$$\beta((S\#T)\#U) = \beta(S\#T) + \beta(U)$$
$$= \beta(S) + \beta(T) + \beta(U)$$
$$= \beta(S) + \beta(T\#U)$$
$$= \beta(S\#(T\#U));$$
$$\omega((S\#T)\#U) = 0$$
$$\iff \omega(S\#T) = \omega(U) = 0$$
$$\iff \omega(S) = \omega(T) = \omega(U) = 0$$
$$\iff \omega(S) = \omega(T\#U) = 0$$
$$\iff \omega(S\#(T\#U)) = 0,$$

so $\omega((S\#T)\#U) = \omega(S\#(T\#U))$.

So the characteristic numbers of the surfaces $(S\#T)\#U$ and $S\#(T\#U)$ are the same. By the Classification Theorem, these surfaces are homeomorphic.

4.2 By Theorem 4.1, we have the following.

(a)

surface	χ	β	ω
S_1	-5	3	0
S_2	-4	0	0
S_3	-5	4	1

(b)

$S_1\#S_1$	-12	6	0
$S_1\#S_2$	-11	3	0
$S_1\#S_3$	-12	7	1
$S_2\#S_2$	-10	0	0
$S_2\#S_3$	-11	4	1
$S_3\#S_3$	-12	8	1

4.3 By Theorem 3.5, the surface defined by the edge equation has characteristic numbers $\chi = 2 - k - n$, $\beta = n$ and $\omega = 1$.

By Theorem 4.1, $\chi(k\mathbb{P}) = 2 - k$ so the surface $k\mathbb{P}\#n\mathbb{D}$ also has characteristic numbers $\chi = 2 - k - n$, $\beta = n$ and $\omega = 1$.

Therefore, by the Classification Theorem, the surfaces are homeomorphic.

4.4 The Klein bottle $\mathbb{K}$ has characteristic numbers $\chi = 0$, $\beta = 0$ and $\omega = 1$. By Theorem 4.1, the surface $2\mathbb{P}$ also has characteristic numbers $\chi = 0$, $\beta = 0$ and $\omega = 1$. Therefore, by the Classification Theorem, $\mathbb{K}$ is homeomorphic to $2\mathbb{P}$.

4.5 Putting $m = 1$, $n = 0$, $k = 1$, $n' = 0$ in Theorem 4.8 gives the result immediately.

4.6 By Theorems 4.2 and 3.3,

$$\chi(S) = 2 - 2m - n, \quad \beta(S) = n, \quad \omega(S) = 0.$$

By Theorems 4.3 and 3.5,

$$\chi(T) = 2 - k - n', \quad \beta(T) = n', \quad \omega(T) = 1.$$

Therefore, by Theorem 4.1,

$$\chi(S\#T) = (2 - 2m - n) + (2 - k - n') - 2$$
$$= 2 - (2m + k) - (n + n'),$$
$$\beta(S\#T) = n + n', \quad \omega(S\#T) = 1.$$

Also, by Theorem 4.1,

$$\chi(U) = \chi((2m + k)\mathbb{P}) + \chi((n + n')\mathbb{D}) - 2$$
$$= (2m + k) - 2(2m + k - 1) + (n + n')$$
$$- 2(n + n' - 1) - 2$$
$$= 2 - (2m + k) - (n + n'),$$
$$\beta(U) = \beta((2m + k)\mathbb{P}) + \beta((n + n')\mathbb{D})$$
$$= 0 + (n + n') = n + n',$$
$$\omega(U) = 1.$$

So $S\#T$ and U have the same characteristic numbers. Hence, by the Classification Theorem, they are homeomorphic.

4.7 For $\chi = -2$, we obtain the following.

$\beta = 0 : 2\mathbb{T}, 4\mathbb{P}$

$\beta = 1 : 3\mathbb{P}\#\mathbb{D}$

$\beta = 2 : \mathbb{T}\#2\mathbb{D}, 2\mathbb{P}\#2\mathbb{D}$

$\beta = 3 : \mathbb{P}\#3\mathbb{D}$

4.8 **(a)** Since $\chi = -3$ is odd, we look for odd values of β such that $\beta \leq 2 - \chi = 5$: these are $\beta = 1$, 3 and 5.

The corresponding surfaces $(1 - \frac{1}{2}(\chi + \beta))\mathbb{T}\#\beta\mathbb{D}$ are $2\mathbb{T}\#\mathbb{D}$ (when $\beta = 1$), $\mathbb{T}\#3\mathbb{D}$ (when $\beta = 3$), and $5\mathbb{D}$ (when $\beta = 5$).

(b) We look for values of β such that $\beta \leq 1 - \chi = 4$: these are $\beta = 0$, 1, 2, 3 and 4. The corresponding surfaces are $5\mathbb{P}$, $4\mathbb{P}\#\mathbb{D}$, $3\mathbb{P}\#2\mathbb{D}$, $2\mathbb{P}\#3\mathbb{D}$ and $\mathbb{P}\#4\mathbb{D}$.

Index